DNA cloning
Volume I

KT-460-335

a practical approach

Edited by
D M Glover

Cancer Research Campaign, Eukaryotic Molecular
Genetics Research Group, Department of Biochemistry,
Imperial College of Science and Technology, London
SW7 2AZ, UK

 **IRL PRESS**
Oxford · Washington DC

IRL Press Limited
P.O. Box 1,
Eynsham,
Oxford OX8 1JJ,
England

First published July 1985
First reprinting February 1986
Second reprinting April 1986

British Library Cataloguing in Publication Data

DNA cloning : a practical approach.—(Practical
  approach series)
  1. Molecular cloning 2. Recombinant DNA
  I. Title II. Series
  574.87′3282   QH442.2

ISBN 0-947946-18-7

Cover illustration. The design for the cover was based on Figure 2 Chapter 2 Volume I, showing a map of λgt11; Figure 10 Chapter 4 Volume II, showing crown gall tumour on *Nicotiana tabacum*; and Figure 4 Chapter 6 Volume II, showing peroxidase stained cells.

Printed in England by Information Printing, Oxford.

Preface

There can be no doubt of the importance of the role played by DNA Cloning techniques in bringing about the information explosion that has occurred in Molecular Biology over the past decade. The use of techniques for recombining DNA *in vitro* is now commonplace in many laboratories and a considerable arsenal of techniques have been developed to tackle a variety of biological problems. Newcomers to the field can get an overview of the weaponry before taking up the pipette by reading texts such as 'Recombinant DNA: A Short Course' by Watson, Tooze and Kurtz (Scientific American Books, New York, 1983); 'Principles of Gene Manipulation' by Old and Primrose (Blackwell, Oxford 1985); or the one by myself 'Gene Cloning: The Mechanics of DNA Manipulation' (Chapman and Hall, London and New York, 1984). The need for the concerted application of the techniques and to keep abreast of the latest technology is a constant challenge. The laboratory manual 'Molecular Cloning' by Maniatis, Fritsch and Sambrook (Cold Spring Harbor Laboratory, New York, 1982) is an invaluable source of protocols for a variety of molecular cloning techniques using that workhorse of Molecular Biologists, *Escherichia coli*. Since the manual was published in 1982, however, there have been considerable developments in *E. coli* host-vector systems and molecular cloning techniques have become well established using other organisms. 'DNA Cloning: A Practical Approach' was conceived as a book that would not duplicate but rather extend and complement the excellent manual of Maniatis and his co-authors. Indeed, throughout the book authors have made constant reference to the Maniatis manual. Such has been the progress of the field that the plan to have a single volume had to be abandoned. The first of the two volumes that have emerged will be concerned primarily with cloning in *E. coli*, whereas alternative prokaryotic and eukaryotic host-vectors are considered in the second volume.

The past three years have seen increased use of bacteriophage lambda vectors that permit the direct selection of recombinants using the Spi⁻ phenotype. Similarly the high efficiency by which recombinant lambda DNAs can be introduced into *E. coli* by *in vitro* packaging has encouraged people to turn to lambda vectors for cDNA cloning. Major technical developments within these two areas are discussed in the first two chapters of volume I. The third chapter presents alternative methods that can be used for cDNA cloning. Vectors that direct the synthesis of fusion proteins have recently come into their own as a means of providing antigenic material in order to raise antibodies against the products of cloned genes. This can of course be turned around, and libraries of DNA can be constructed in these vectors for screening using available antibodies as probes. Some of the bacteriophage lambda vectors described in Chapter 2 are suitable for these purposes as are one set of such plasmid vectors described in Chapter 4. Another imaginative group of vectors, the pEMBL plasmids, is described in Chapter 5. These vectors can be propagated for physical mapping experiments as any other plasmid in the form of double-stranded DNA. If propagated in bacteria carrying F-factors, however, and then super-infected with the male-specific phage f1, they generate single-stranded DNA that is packaged into virions and which can be used for DNA sequencing or *in vitro* mutagenesis. Two approaches to the mutagenesis *in vitro* of DNA carried in 'single-stranded vectors' are described in Chapters 7 and 8. The attractions of the high *in vitro* packaging efficiencies possible when using bacteriophage lambda vectors has not stifled the 'opposition' from those who favour plasmid vectors for particular cloning purposes. On the contrary, methods have been developed to achieve high efficiency transformation of *E. coli* with naked DNA and these are presented in Chapter 6. Finally, Chapter 9 of the first volume looks at a set of vectors with a broader host range which allow the propagation of plasmid recombinants within Gram negative bacteria other than *E. coli*.

In the second volume, alternative bacterial hosts are explored, including *Bacilli* and *Steptomycetes*. The remaining part of the volume focusses on eukaryotic systems with chapters describing transformation system in yeast, *Drosophila* and in plants. The field of animal cell host-vector systems is rapidly evolving, and it may be that in a short while there may be a need for a third volume to cover this area more thoroughly. Meanwhile, the general procedures for introducing DNA into animal cells are covered and two viral vector systems are also described.

I hope that the community of Molecular Biologists will find these volumes useful as I am sure this will give the authors the best gratification. My own thanks go to all the authors for their willing participation in this project, for producing their manuscripts so promptly, and for being so tolerant of the interfering editor.

David M. Glover

Contributors

C.Baldari
European Molecular Biology Laboratory, Postfach 10.2209, 6900 Heidelberg, FRG

G.Cesareni
European Molecular Biology Laboratory, Postfach 10.2209, 6900 Heidelberg, FRG

G.Ciliberto
European Molecular Biology Laboratory, Postfach 10.2209, 6900 Heidelberg, FRG

R.Cortese
European Molecular Biology Laboratory, Postfach 10.2209, 6900 Heidelberg, FRG

R.W.Davis
Department of Biochemistry, Stanford University School of Medicine, Stanford, CA 94305, USA

L.Dente
European Molecular Biology Laboratory, Postfach 10.2209, 6900 Heidelberg, FRG

F.C.H.Franklin
Department of Genetics, The University of Birmingham, P.O. Box 363, Birmingham B15 2TT, UK

H.-J.Fritz
Max-Planck-Institut für Biochemie, Abteilung Zellbiologie, Am Klopferspitz, D-8033 Martinsried bei München, FRG

H.-W.Griesser
Institut für Genetik, Universität zu Köln, D-5000 Köln, FRG

D.Hanahan
Cold Spring Harbor Laboratory, Cold Spring Harbor, NY 11724, USA

T.V.Huynh
Department of Biochemistry, Stanford University School of Medicine, Stanford, CA 94305, USA

B.Müller-Hill
Institut für Genetik, Universität zu Köln, D-5000 Köln, FRG

J.F.Jackson
Department of Biochemistry, Imperial College of Science and Technology, Imperial College Road, London SW7 2AZ, UK

K.Kaiser
University of Glasgow, Institute of Genetics, Church Street, Glasgow G11 5JS, UK

M.Koenen
Institut für Genetik, Universität zu Köln, D-5000 Köln, FRG

N.E.Murray
University of Edinburgh, Department of Molecular Biology, King's Buildings, Mayfield Road, Edinburgh EH9 3JR, UK

M.Sollazzo
European Molecular Biology Laboratory, Postfach 10.2209, 6900 Heidelberg, FRG

C.Traboni
Istituto di Scienze Biochemiche, University of Naples, Naples, Italy

C.J.Watson
Cancer Research Campaign Eukaryotic Molecular Genetics Research Group, Department of Biochemistry, Imperial College of Science and Technology, Imperial College Road, London SW7 2AZ, UK

R.A.Young
The Whitehead Institute for Biomedical Research and Department of Biology, Massachusetts Institute of Technology, Cambridge, MA 02139, USA

Contents

2. CONSTRUCTING AND SCREENING cDNA LIBRARIES IN λgt10 AND λgt11

T.V.Huynh, R.A.Young and R.W.Davis

Abbreviations

APRT	adenine phosphoribosyl transferase
ars	autonomously replicating segment
BPV-1	bovine papillomavirus type 1
BUdR	5-bromodeoxyuridine
C23O	catechol 2,3-oxygenase
CaMV	cauliflower mosaic virus
CAT	chloramphenicol acetyltransferase
CEF	chick embryo fibroblasts
CHO	Chinese hamster ovary
Cm	chloramphenicol
CRM	cross-reacting material
CTAB	cetyl triethylammonium bromide
DHFR	dihydrofolate reductase
DMEM	Dulbecco's modified Eagle's medium
DMSO	dimethylsulphoxide
d.s.	double-stranded
DTT	dithiothreitol
EtBr	ethidium bromide
FCS	foetal calf serum
βgal	β-galactosidase
gdDNA	gapped duplex DNA
GH	growth hormone
HA	haemagglutinin
HBS	Hepes-buffered saline
HbsAg	hepatitis virus surface antigen
HGT	high gelling temperature
HSV-1	herpes simplex virus type 1
IF	interferon
IMP	inosine monophosphate
IPTG	isopropyl-1-thio-β-D-galactoside
Kn	kanamycin
LGT	low gelling temperature
LTR	long terminal repeat
2ME	2-mercaptoethanol
MEM	Eagle's minimal medium
MES	2[N-morphino]ethone sulphonic acid
MMTV	mouse mammary tumour virus
MSV	Moloney murine sarcoma virus
NPTII	neomycin phosphotransferase
NRDC	National Research and Development Corporation
ONPG	o-nitrophenyl-β-D-galactopyranoside
ORF	open reading frame
PBS	phosphate-buffered saline
PEG	polyethylene glycol
RNP	ribonucleoprotein
RSV	Rous sarcoma virus
SAM	S-adenosyl-L-[methyl-^{3}H]methionine

SB	simple transformation buffer
SDS	sodium dodecylsulphate
Sm	streptomycin
s.s.	single-stranded
SSC	standard saline citrate
Su	sulphonamides
T-DNA	transforming DNA of the Ti-plasmid
TBS	Tris-buffered saline
Tc	tetracycline
TES	N-tris(hydroxymethyl)methyl-2-aminoethanesulphonic acid
TFB	transformation buffer
Ti plasmid	tumour inducing plasmid
TK	thymidine kinase
X-gal	5-bromo-4-chloro-3-indolyl-β-D-galactoside

CHAPTER 1

The use of Phage Lambda Replacement Vectors in the Construction of Representative Genomic DNA Libraries

KIM KAISER and NOREEN E. MURRAY

1. THE PERFECT LIBRARY

The perfect genomic DNA library would contain DNA sequences representative of an entire genome, in a stable form, as a manageable number of overlapping clones. The cloned fragments would be large enough to contain whole genes and their flanking sequences. On the other hand they should be small enough to be mapped easily by restriction enzyme analysis. Most important, a library should be both easy to construct from small amounts of starting material, and easy to screen for the sequence of interest, usually by hybridisation with radioactive DNA or RNA probes. Overlapping clones facilitate 'walking' into adjacent regions of the genome, and to make this simple it should be possible to excise the inserted DNA free of vector sequences so that it in turn can be used as a radioactive probe. Ideally, it should be possible to amplify the library without loss or misrepresentation, and to store it for years without significant decrease in titer.

At present only phage lambda and cosmid vectors fulfil, even in part, these requirements. Cosmid vectors have a larger capacity (~ 45 kb) but are generally more difficult to use than phage lambda vectors, which restrict the insert length to a maximum of $20-25$ kb. In either case fairly large libraries ($10^5 - 10^6$ individual recombinants) are likely to be required to represent a complex eukaryotic genome. Currently, phage lambda vectors provide the easier and more efficient means for the routine construction of genomic DNA libraries (see Section 17 for a discussion of the relative merits of cosmids and lambda).

2. THE PRINCIPLES OF CLONING IN LAMBDA REPLACEMENT VECTORS

The judicious choice and use of lambda vectors really requires appreciable understanding of the interaction of phage λ with its host. However, we have chosen to introduce the basic technicalities of the construction of representative libraries with minimal reference to the biology of phage λ, and only then discuss in detail those aspects of phage λ and its host (Section 13) that are relevant to a more informed use of λ vectors.

The chromosome of bacteriophage lambda is a linear DNA molecule 48.6 kb long (3) but nearly 40% of the genome is inessential for the propagation of the phage. As the result of many *in vitro* and *in vivo* manipulations of the lambda

1

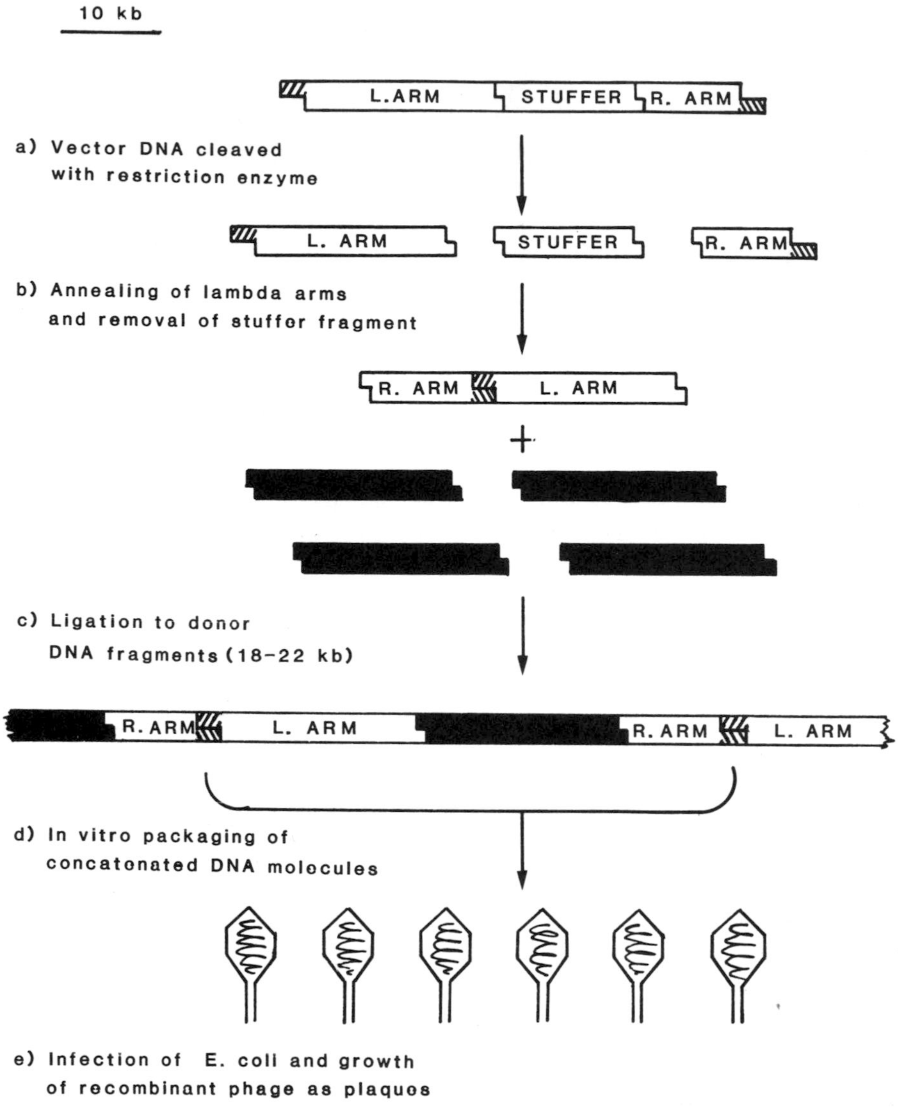

10 kb

	L.ARM	STUFFER	R. ARM

a) Vector DNA cleaved
   with restriction enzyme

L. ARM   STUFFER   R. ARM

b) Annealing of lambda arms
   and removal of stuffer fragment

R. ARM   L. ARM

+

c) Ligation to donor
   DNA fragments (18–22 kb)

R.ARM   L. ARM   R. ARM   L. ARM

d) In vitro packaging of
   concatenated DNA molecules

e) Infection of E. coli and growth
   of recombinant phage as plaques

Figure 1. The basic strategy of cloning using a lambda replacement vector. Cleavage of the vector DNA with the appropriate enzyme generates cohesive termini to which are ligated donor DNA fragments bearing complementary termini. Hatched boxes denote the lambda *cohesive ends*. As shown here the *cohesive ends* were allowed to anneal before ligation. The concatenated ligation product is packaged *in vitro* into infective phage particles, each of which is recovered in an amplified form as a plaque on the surface of an agar plate. A replica of the plaque distribution can be obtained by 'blotting' the surface of the plate with a nitrocellulose filter (Section 9 and 18.10), to which a probe DNA may then be hybridised.

genome, we now have a bewildering variety of vectors for the propagation of non-lambda DNA (1,2,4,5,6,7,8). Here, we shall be concerned only with the class of vectors known as replacement vectors, particularly those that provide maximum space for exogenous DNA fragments, thereby reducing the number of individual recombinants necessary for 'complete' representation of a whole genome.

The strategy for the use of any replacement vector is shown in *Figure 1*. Cleavage of the vector DNA with the appropriate restriction enzyme generates three fragments (*Figure 1a*): a left arm and a right arm, which together contain all the information essential for the production of infective phage particles, and a dispensable central fragment (sometimes called a 'stuffer' fragment). This central region contains no essential genes and serves merely to bolster up the size of the vector DNA. Replacement vectors are so called because their central fragment of DNA may be exchanged for an exogenous DNA fragment. Only genomes in the 40 – 52 kb range are packaged into phage particles with high efficiency. Ligation of the two arms of a replacement vector generates a DNA molecule that is too small to be packaged. The requirement for a genome that is longer than the two arms of the vector thus provides a powerful enrichment for recombinant genomes.

The DNA ends generated by cleavage with the restriction enzyme have self-complementary cohesive termini (*Figure 1a*). If the cleaved vector DNA is mixed with exogenous DNA fragments that have the same cohesive termini, and DNA ligase is added, a proportion of the novel DNA molecules will contain an exogenous DNA fragment flanked by left and right arms. When introduced into a bacterial cell, such a DNA molecule will enter the lytic growth cycle and produce a clone of viral particles. Individual plaques of viral growth originating from a singly infected bacterium are, therefore, a source of amplified recombinant chromosomes containing a specific sequence from the population of donor fragments.

This simple picture is complicated somewhat by the fact that the vector DNA molecule itself possesses complementary single-stranded projections at each end of the molecule (*Figure 1a*). These *cohesive ends* allow the left and right arms to anneal with each other (*Figure 1b*), and permit the formation of chain-like (concatenated) recombinant DNA molecules (*Figure 1c*). Monomeric genomes, however, can be readily recovered from concatenated DNA (Section 4.1 and 13.1).

Phage DNA molecules, recombinant or otherwise, can be introduced into a bacterial cell as naked DNA (*transfection*) in which case efficiencies of $> 10^4$ recombinant clones per μg of donor DNA can be obtained. This efficiency would suffice for the construction and propagation of libraries from small genomes (e.g., *E. coli* or yeast). Considerably higher efficiencies are preferable, however, when working with larger genomes in order that the starting material should not be a limiting factor in library construction, and that recombinant plaques can be obtained at high density. Greater than 10^6 recombinant clones per μg of donor DNA can be obtained by '*in vitro* packaging' of the recombinant DNA molecules into phage particles before introduction to the bacteria (*Figure 1d*).

To generate a 'complete' library it is now usual to fragment high molecular weight genomic DNA in a pseudo-random manner by partial cleavage with a restriction enzyme that cuts frequently compared with the desired insert size, and that also generates cohesive termini appropriate to the vector. By such means an overlapping set of DNA fragments, representative to a first approximation of the entire genome, is obtained. Only exogenous DNA fragments that produce a phage DNA molecule in the packagable size range will be present in the clones. This requirement can be achieved by insertion of either a single large DNA fragment, or several smaller fragments (multiple insertion). Multiple inserts are not desirable since the juxtaposition of DNA fragments that are normally non-contiguous in the genome creates problems in the analysis of the recombinant DNA molecules. It is common, therefore, to fractionate the exogenous DNA fragments so that only those of desirably large size are included in the ligation reaction. Multiple insertion will then produce over-sized phage genomes which cannot be packaged.

Unless the normal central fragment is removed from the cleaved vector DNA before ligation with exogenous DNA fragments, a proportion of the newly formed phage DNA molecules will have re-assimilated this fragment. Such events can be prevented by physical separation of the central fragment from the vector arms before ligation (*Figure 1b*). Every viable phage genome produced by ligation should then contain exogenous DNA, and the library will, in theory, be uncontaminated with parental sequence phage. Purification of the vector arms is not a stringent requirement, however, in the case of vectors that allow for a genetic selection against non-recombinant phage during propagation of the library (Sections 14 and 15).

In the sections to follow we shall have cause to elaborate considerably on the simplified scheme outlined above. Since many of the relevant methods have been adequately described elsewhere, we will often refer the reader to these sources, notably the ubiquitous Molecular Cloning: A Laboratory Manual (2), and the Experimental Methods section of Lambda II (1). In this chapter we pay special attention to stages in the cloning procedure where pitfalls commonly occur, and suggest ways in which these can be overcome. In particular we stress the importance of including, wherever possible, control experiments that serve as a guide to the success or otherwise of each step of the procedure.

3. DONOR AND VECTOR DNAS

3.1 Isolation of Donor DNA

The donor DNA, whatever the source, must fulfil one essential requirement: that it be of very high molecular weight. Some degree of breakage is inevitable during the isolation of genomic DNA from cells, due to mechanical shearing during the isolation process, and to the action of degradative enzymes released during rupture of the cells. Breakage must be kept to a minimum, particularly when the random collection of donor DNA fragments is to be generated by partial digestion with a restriction enzyme since otherwise a high proportion of the fragments will have non-cohesive ends. Such fragments inhibit the cloning procedure by joining non-productively to vector DNA fragments, since only one of the phage arms, both of which are essential to the production of an infective recombinant DNA molecule, can attach to them. Moreover, concatenated phage DNA molecules are the usual substrate for *in vitro* packaging of the DNA into infective phage particles (*Figure 1c* and Section 4.1). Ragged donor fragments acting as 'chain-stoppers' will thereby inhibit both the ligation and the packaging reactions.

These considerations will be less important when working with a small genome (e.g., that of *E. coli* which can be adequately represented in a library of less than 1000 individual recombinants having inserts of 20 kb in length) than with a complex eukaryotic genome for which a million or so recombinants may be required. They are always important, however, when supplies of donor DNA are limiting. As a rule of thumb the average size of the donor DNA fragments must be considerably greater than 100 kb if high efficiencies of recovery are to be obtained.

The ideal isolation procedure would contain the minimum number of steps, and would completely eliminate mechanical shearing or enzymatic degradation. A common source of mechanical shearing is over-zealous mixing of viscous DNA solutions, especially when protocols that demand several cycles of extraction with phenol, or other organic solvents, are followed. Most procedures for the isolation of high molecular weight DNA require the suspension of spheroplasts (prokaryotes) or eukaryotic nuclei in a stabilising buffer, at which stage a lysing agent is introduced which releases the DNA into solution. The use of strong detergents as lysing agents not only releases DNA but simultaneously acts as an inhibitor of endogenous DNase activity.

Before the lysis step, mechanical methods (e.g., homogenisation) are permissible. After lysis the solution becomes viscous due to the presence of high molecular weight DNA. From this stage onwards it is essential that mechanical shearing (mixing, shaking, extrusion through pipettes or syringe needles) be kept to an absolute minimum, and that the number of steps likely to cause shearing be reduced as much as possible. In particular, mixing during phenol extraction should be carried out *gently*! (Section 18.4). Since extremely concentrated, and hence viscous, DNA solutions are particularly difficult to manipulate, it is usually better to work with moderate quantities of DNA (< 1 mg/20 ml of nuclear lysate) and to choose the amount of starting material accordingly.

A method that avoids phenol extraction and ethanol precipitation, and which involves simultaneous concentration and purification by equilibrium centrifugation in caesium chloride (CsCl) gradients, is described in Section 18.4. This method, which has been adapted from the one described in reference 27, has been used successfully to isolate high molecular weight DNA from bacteria, fungi, *Drosophila*, cultured mammalian cells, and solid mammalian tissue. It is fast, involves a minimum number of steps, and may be followed easily by workers having little experience in the handling of high molecular weight DNA.

3.2 Fractionation of Donor DNA

A truly random library can only be obtained from DNA that has been fractionated in a sequence-independent manner, for example by controlled mechanical shearing (9). It is usually more convenient, however, to fractionate DNA by partial digestion with restriction endonucleases. If the enzymes used cut relatively frequently compared with the desired fragment size, partial cleavage will produce a set of overlapping fragments which is sufficiently random for most purposes. It cannot be guaranteed, however, that all regions of the genome are represented in the desired size range since some large regions with no cleavage sites may

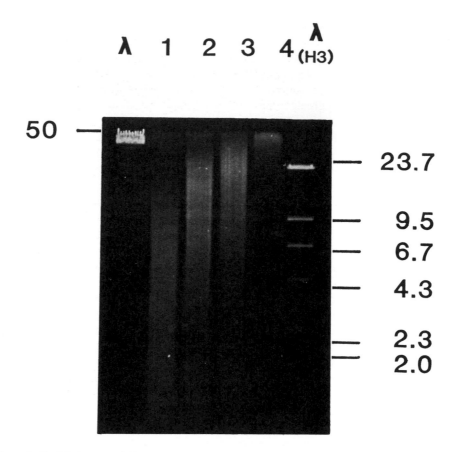

Figure 2. Partial cleavage of high molecular weight DNA with *Sau*3A. A series of trial reactions was carried out as described in Section 18.5. DNA digested at concentrations of enzyme that embrace the optimum cleavage conditions were separated by electrophoresis overnight at 20 V in a 0.3% agarose gel. The figure is a photograph of the gel after staining with ethidium bromide. The markers are intact lambda DNA and lambda DNA that has been cleaved to completion with *Hind*III. The lambda DNA samples were heated to 68°C for 10 min, before loading, in order to disassociate their cohesive ends. In the large scale reaction, 200 µg of high molecular weight DNA was digested, half under the conditions of lane 2 and half under the conditions of lane 3. Note that a portion of the DNA is relatively refractory to cleavage by *Sau*3. Migrating, as it does, behind the intact lambda DNA, this portion cannot be cloned using either a lambda or a cosmid vector.

exist. Moreover, all restriction targets may not be cleaved with equal efficiency.

Because of their compatibility with *Bam*HI (Section 3.3), the enzymes *Sau*3A and *Mbo*I are used most often. Both enzymes recognise a tetra-nucleotide DNA sequence (GATC) that will occur, on average, once every 4^4 (256) base pairs in random sequence DNA. Thus, to generate 20 kb fragments, one need cut only 1/80 of the available cleavage sites. This ratio will vary with the GC content of the donor DNA used, however, so it is usual to determine the optimum enzyme concentration empirically in a trial reaction (*Figure 2* and Section 18.5).

Under optimum conditions, the size distribution of fragments will peak in the most desirable range (18−22 kb for the vectors described here). In such a prepara-

high salt   low salt

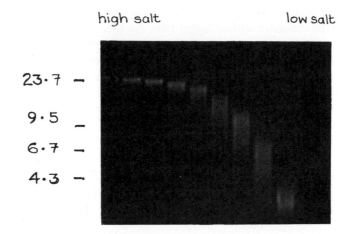

23·7 —

9·5 —

6·7 —

4·3 —

Figure 3. Electrophoretic analysis of donor DNA partially cleaved with *Sau*3A and fractionated by velocity gradient sedimentation. Partially digested DNA is fractionated as described in Sections 3.2 and 18.6 and aliquots electrophoresed on a 0.3% agarose gel as described in the legend to *Figure 2*.

tion there will, nevertheless, be a proportion of small fragments which, if not removed, can become incorporated into the vector as multiple inserts. Large fragments, on the other hand, will give rise to oversized recombinant molecules that cannot be packaged. Moreover, recombinant DNA molecules just at the upper and lower limits of the packagable size-range may be unduly susceptible to Rec-dependent deletion and duplication, or Rec-independent deletion and rearrangement (Section 16.2). Fragments in the desirable range can be separated physically from the rest either by velocity centrifugation through sodium chloride or sucrose gradients (Section 18.6), or by electrophoresis in agarose gels (7). The former procedure generally provides the most suitable DNA since agarose or other contaminants inhibitory to ligase are sometimes difficult to remove from DNA recovered from gels. The range of molecular lengths present in each salt/sucrose gradient fraction is determined by agarose gel (0.3%) electrophoresis (see *Figure 3*), and fractions that embrace the most desirable insert size are retained for further testing. The quantity of DNA in each fraction is determined by visual estimation of fluorescence due to the intercalating dye, ethidium bromide (Section 18.6). DNA in each gradient fraction will have to be concentrated, usually by ethanol precipitation, before it can be used for cloning. Indeterminate losses can occur when small amounts of DNA are ethanol precipitated (Section 18.2), so it is sensible to determine the amount of DNA in each fraction *after* it has been concentrated. Alternatively the partially cleaved DNA may be treated with phosphatase (see Section 18.3 for notes on checking phosphatase preparations) and ligated directly to the cleaved vector DNA. Ligase cannot join two DNA molecules if *both* are lacking a 5′ terminal phosphate group. Phosphatase-treated donor DNA fragments are therefore unable to form multiple inserts. There will still be a proportion of recombinant DNA molecules that are too small or too large to be packaged and it is left to the *in vitro* packaging extracts (Section 4.2) to discriminate against them. Even so, insert size will vary considerably (the vectors described here accept fragments ranging from 9 to 22 kb) so that a representative library constructed in this manner will be larger than the theoretical minimum, especially if packaging extracts which do not impose a strong bias against small recombinant DNA molecules are used. It should also be noted that recombinant DNA molecules just at the upper and lower limits of the packagable size-range, and which are potentially unstable, will be produced by this method. Their recovery will be avoided to some extent, at least at the lower limit, by an appropriate choice of packaging extracts and buffers (Section 4.2).

3.3 Vector DNA

The structure of a typical replacement vector DNA molecule (EMBL3) is shown in *Figure 4a*. A central fragment is flanked by the two essential arms, the arm on the left encoding the virion proteins, and the arm on the right the origin of

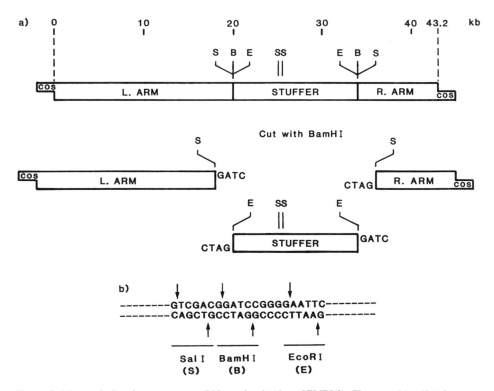

Figure 4. (a) A typical replacement vector DNA molecule (that of EMBL3). The central 'stuffer' fragment is flanked by polylinkers (**b**) in inverse orientation. Other than two *Sal*I sites in the central fragment, the *Sal*I (S), *Bam*HI (B), and *Eco*RI (E) sites in the polylinkers are the only sites for these enzymes in the entire DNA molecule. Hence EMBL3 may be used for the cloning of DNA fragments with termini complementary to those generated by any of these enzymes. The 4 base long single stranded projections generated by *Bam*HI cleavage (GATC) allow the arms to be ligated to donor DNA fragments bearing termini created by partial cleavage with *Sau*3A. *Sal*I sites retained in the arms permit the inserts of recombinant DNA molecules to be excised with minimum contamination by vector DNA sequences. *Eco*RI sites retained in the central fragment allow it to be enzymatically isolated before ligation to donor DNA fragments. *cos* denotes a lambda *cohesive end*. (**b**) The DNA sequence of the relevant region of an EMBL3 polylinker.

8

replication, the major promoters and a number of other essential genes. The DNA is shown in the linear form as isolated from phage particles. Both arms carry at their extremities short (12 base) complementary single-stranded projections. These *cohesive ends (cos)* permit the λ genome to circularise following infection and are regenerated when the phage genomes are packaged.

The central fragment of EMBL3 is flanked by polylinkers (*Figure 4b*) in inverse orientation:

*Sal*I-*Bam*HI-*Eco*RI−central fragment−*Eco*RI-*Bam*HI-*Sal*I

These polylinkers, which are less than 40 bp long, include the only sites for these three enzymes in the vector (except for two additional *Sal*I sites in the central fragment), so any of these three enzymes can be used for cloning. The vector was designed, however, especially for use with *Bam*HI which makes a staggered break at the hexanucleotide sequence:

$$\stackrel{\downarrow}{G}GATCC$$
$$CCTAG\stackrel{\uparrow}{G}$$

Any hexanucleotide sequence occurs, on average, only once every 4^6 (4096) base pairs in random sequence DNA, rather too seldom for the 'random' fractionation of donor DNA. The core of the *Bam*HI recognition sequence, however,

$$\stackrel{\downarrow}{G}ATC$$
$$CTAG\stackrel{\uparrow}{}$$

is the recognition site of the enzymes *Sau*3A and *Mbo*I which cleave it as shown. Thus, EMBL3 DNA cleaved by *Bam*HI can accept an exogenous DNA fragment prepared by partial digestion with *Sau*3A or *Mbo*I (Sections 3.2 and 18.5). The inserts of recombinant derivatives of EMBL3 are then flanked by *Sal*I sites that enable them to be excised cleanly from the vector arms, and used as probes to recover overlapping, or related, recombinant phages from the library (Section 12).

Although the linkers may differ, similar considerations apply to the other vectors described here (*Figure 8*). All contain *Bam*HI sites that are closely flanked on their outsides by sites for other commonly available restriction enzymes. Furthermore, all can be used for cloning donor fragments generated by cleavage with other enzymes (e.g. *Eco*RI and *Sal*I in the case of EMBL3). This may be appropriate if a desired sequence is known to be contained within an *Eco*RI or *Sal*I fragment that lies in the acceptable size range (9−22 kb) but a fully representative, and therefore more generally used, library cannot be constructed in this manner.

The quality of the vector DNA, while somewhat easier to control, is of no less importance than that of the donor DNA. Several methods for preparing phage particles on a large scale (≥ 100 μg) have been published (see references 1 and 2). Just as with the preparation of donor DNA, it is important not to be too ambitious. Excessive amounts of starting material will produce concentrated, and hence viscous, DNA which is difficult to work with and which may be somewhat refractory to enzymatic cleavage, even after extensive phenol extraction. The purest phage preparations are obtained by double banding in equilibrium CsCl gradients. After removal of CsCl by dialysis, the phage are diluted appropriately and gently extracted several times with either phenol or a phenol:chloroform:iso-amyl alcohol mixture (25:24:1) to release the phage DNA (6). Phenol should be removed by extensive dialysis and not by ethanol precipitation and ethanol washing, as a con-catenated precipitate is difficult to resuspend. Great care should be taken to maintain the sterility of phage DNA solutions as the single-stranded cohesive ends are extremely vulnerable and their integrity is essential.

The vector DNA should be tested in several ways before use in cloning. Particular attention should be paid to the points listed in Section 18.7.

3.4 **Removal of the Central Fragment**

Where a genetic selection against non-recombinant phages is unavailable, or undesirable, the central fragment can be effectively removed from the cleaved vector DNA population before ligation to donor DNA fragments, thereby ensuring that the vast majority of viable phage genomes produced are recombinant.

Physical separation of the vector arms from the central fragment can be achieved by sodium chloride or sucrose gradient fractionation, or by elution or extraction from an agarose gel. For the reasons mentioned in Section 3.2, we prefer the former method. In either case care should be taken after cleavage to ensure that all of the vector arms have annealed via their *cohesive ends* (*Figure 1* and Section 18.8). It is then only necessary to isolate a single DNA species (a composite of left and right arms), which by virtue of its large size can be easily resolved from the central fragment. A procedure for the purification of phage arms is found in reference 2.

An alternative method of separation has been used for EMBL3 (6). The *Bam*HI sites of this vector are flanked internally by *Eco*RI sites (see *Figure 4*). Cleavage with both *Bam*HI and *Eco*RI releases short (<10 bp) *Bam*HI-*Eco*RI fragments from both ends of the central fragment. The central fragment is thereby rendered unable to compete with donor DNA fragments for ligation to the vector arms. Moreover if isopropanol, rather than ethanol, is used to precipitate the vector DNA, the linker fragments are sufficiently small that they remain in solution (see Section 18.8). This double digestion method is not generally applicable, however, because many vectors do not have appropriate restriction sites juxtaposed. Furthermore, even if two enzymes are used simultaneously, in which case their buffer and co-factor requirements must be compatible, any given polylinker sequence will be cleaved initially by one enzyme, and the second enzyme must then be able to cleave at a site extremely close to the end of a DNA molecule (less than 10 bp away in the case cited above). Some enzymes, apparently including certain batches of *Sal*I, may cut terminal sites with reduced efficiency. For any pair of enzymes the suitability of this method must be determined by trial and error (but see reference 10). For EMBL3 vector DNA, first cut with *Bam*HI then, when cleavage is complete, adjust the reaction buffer to *Eco*RI conditions and digest with a 3-fold excess of *Eco*RI.

Whether removal of the central fragment is by physical or enzymatic methods, it is important to assay the 'background' of phages produced when the vector DNA preparation is ligated to itself. Do not assume this to be zero. For special applications it may be necessary to reduce background to the absolute minimum. For EMBL3 and EMBL4, one can both remove (or inactivate) the central fragment *and* impose a genetic selection against parental sequence phage (Sections 14 and 15).

4. COMBINATION AND PACKAGING

For the most part we shall assume that infective phage particles are to be produced by *in vitro* packaging of recombinant DNA molecules which, as described above, is the most effective way of introducing phage DNA into bacterial cells. Transfection, as a less efficient but cheaper alternative to *in vitro* packaging, is discussed in Section 4.3.

Greater than 10^8 infective phage per μg of lambda DNA can be obtained by *in vitro* packaging. This efficiency allows more than 10^6 recombinant DNA molecules to be recovered per μg of donor DNA. *In vitro* packaging is carried out in 'packaging extracts', which are prepared following the induction of *E. coli* lysogenic for certain mutant derivatives of lambda. The extracts contain all of the structural proteins necessary for the assembly of lambda DNA into an infective phage particle. They are available commercially and can be stored at $-70°$C until required. A description of their preparation is beyond the scope of this article and the reader is advised to consult references 1 and 2.

4.1 The Ligation Reaction

Having satisfied oneself as to the suitability of a particular batch of cleaved vector DNA, which if prepared on a $20-50$ μg scale will suffice for the testing and construction of several libraries, it is necessary to determine the optimum concentrations of vector and donor DNAs to be included in the ligation reaction. Both concatenated and, perhaps to a somewhat lesser extent, linear lambda DNA molecules are efficient substrates for *in vitro* packaging, whereas monomeric circles are not. Ligation conditions chosen as described here (Section 18.9) will tend to favour the production of long concatenated DNA molecules. This requires that the vector arms be annealed before inclusion in the reaction (*Figure 1*), and that the reaction be carried out at relatively high DNA concentration (discussed in Section 4.3).

In theory (2) it should be possible to calculate, *a priori*, the concentrations of the participating DNA molecules that will maximise the formation of the desired ligation product. In practice a proportion of the participating molecules will have damaged ends or other problems and, although the concentration of vector DNA may be known relatively precisely, the amounts of donor DNA available will have been only roughly determined by visual estimation of ethidium fluorescence. This is not the most accurate of methods. In view of these complications, and if donor DNA is not a limiting factor, it is often more convenient to determine the optimum conditions empirically, although one may be guided in this by the theoretical values. The simplest approach is to analyse each variable independently in a series of test ligations as outlined in Section 18.9.

4.2 The Packaging Reaction

There are both lower ($\sim$40 kb) and upper ($\sim$52 kb) limits to the size of phage DNA molecules that can be packaged into phage heads (1). By preparing packaging extracts in different ways, however, it is possible to vary the efficiency with which small genomes ($\sim$40 kb) are recovered relative to genomes of normal size (2,11). Protocol II of reference 2 imposes a fairly stringent selection for DNA molecules close in size to that of wild-type lambda DNA (i.e., $\sim$20 kb inserts if the vector arms are 30 kb). This protocol is useful in order to ensure that inserts are of the largest possible size, particularly if no size-fractionation step is included in the preparation of the donor DNA fragments. A more catholic range of inserts, down to the minimum acceptable to the vector, can be obtained by using protocol I of reference 2 (see also the discussion on page 298 of this reference). This may be useful if the maximum yield of recombinants is to be obtained from limiting amounts of cleaved donor DNA, with the proviso that protocol I may be inherently less efficient than protocol II. Most packaging extracts obtained from commercial sources will be of the high stringency form.

4.3 Transfection

Packaging extracts are costly if obtained commercially. It is possible to generate libraries of prokaryotic genomes by *transfection* (the uptake of naked phage DNA by bacteria). The desirable substrate for transfection is a monomeric phage DNA molecule. Optimum ligation conditions for transfection will not, therefore, be the same as those determined as optimal for *in vitro* packaging.

Even without regard to DNA concentration, the choice of ligation conditions always involves a compromise. The Tm for cohesive termini generated by restriction enzymes (4 bases) is around 5°C. (The Tm is the temperature at which half the complementary nucleic acid species in a solution are annealed.) Commonly used DNA ligases, on the other hand, work best at 37°C and have relatively little activity at 5°C. A reaction temperature of $10-16$°C is often chosen and, with ligase in considerable excess, the reaction proceeds at a reasonable rate (minutes to hours). The annealing of the lambda *cohesive ends*, which have a higher activation energy, will still be far from equilibrium after a short reaction at low temperature. Thus, if the arms are deliberately disassociated before inclusion in the reaction (by heating at 68°C for 10 min and cooling rapidly on ice), the predominant ligation products will be linear monomeric recombinant DNA molecules suitable for transfection (11,15). Alternatively, to maximise the production of concatenated DNA molecules, the arms should be pre-annealed (Section 18.9). In either case the ligation reaction is carried out at high DNA concentration in order that inter-molecular joining is favoured over intra-molecular joining (circularisation).

5. SCALING UP

Having determined the optimum ligation conditions and taken into account background, set up a large-scale reaction that will generate at least the number of recombinant phages needed for a 'complete' library (Section 6), and package the ligated DNA in extracts of the highest efficiency. Regretably, the efficiencies obtained on a small scale will seldom be achieved at this crucial stage of the operation, so it is inadvisable to commit ones entire stock of precious fractionated DNA to one experiment.

For example, the efficiency with which batches of packaging extracts package phage DNA molecules may not be strictly linear with respect to DNA concentration, especially at high concentrations. If the contents of several vials of extract are needed they should be used separately, as pooling them generally leads to reduced efficiency (J. Wolfe, personal communication). Only work with two or three vials at once so that the extracts are incompletely thawed when mixed with the DNA.

Always titer the packaged phage, after adding a drop of chloroform to the tube, and store the library at 4°C until required. Packaging extracts can inhibit bacterial growth if present in excess. Should the titer be so low that greater than one tenth to one quarter of a packaging reaction must be applied to a single 90 mm culture dish in order to obtain enough plaques for screening (Section 9) or amplification (Section 8), concentrate the packaged phage by centrifugation in a CsCl gradient (1,2). Conditions for the long term storage of lambda are described in references 1, 2 and 15, although it seems that phage packaged *in vitro* have a shorter half-life than those obtained by growth *in vivo*.

In spite of these problems, lambda libraries of $> 10^6$ independent recombinant phages can be obtained routinely from small amounts of starting material although, in view of the last point, it is preferable to screen or amplify them as quickly as possible.

6. WHAT SIZE OF LIBRARY?

Assuming complete randomness of sequence representation, if each inserted DNA fragment is of identical size, and if the genome size is known, then it is a simple matter to determine the size of library that will have an arbitrary probability of including a particular DNA sequence. In practice, although none of these conditions is strictly true, it is possible to arrive at an approximation sufficient for most purposes.

If x is the insert size and y is the size of the haploid genome (in the same units), then a library of

$$N = \frac{\ln(1-p)}{\ln(1-x/y)}$$

clones will have a probability p of containing any particular DNA sequence (12). Assuming 20 kb inserts and making p = 0.99 we obtain the library sizes in *Table 1*. For comparison we show the sizes of library that would be necessary if 45 kb fragments were cloned using a cosmid vector. It should be noted that non-haploid organisms will require larger libraries for complete representation if heterozygous sequences are to be cloned.

Table 1. The Number (N) of Independent Recombinant DNA Molecules that must be Recovered in either a Lambda Replacement Vector, or a Cosmid, in order to have a 99% Probability of having Cloned a Particular DNA Sequence.

Organism	Approximate DNA content of haploid genome (bp)	N for 20 kb inserts[a]	N for 45 kb inserts[a]
Escherichia coli (bacterium)	4.2×10^6	9.2×10^2	4.3×10^2
Saccharomyces cerevisiae (yeast)	1.4×10^7	3.2×10^3	1.4×10^3
Drosophila melanogaster (fruit fly)	1.4×10^8	3.2×10^4	1.4×10^4
Stronglyocentrotus purpuratus (sea urchin)	8.6×10^8	2.0×10^5	8.8×10^4
Homo sapiens (man)	3.3×10^9	7.6×10^5	3.4×10^5
Triticum aestivum (hexaploid wheat)	1.7×10^{10}	3.9×10^6	1.7×10^6

[a]The calculations were based on the assumption that cleavage of the donor DNA occurs at random, which will not be strictly true for DNA cleaved by partial digestion with *Sau*3A or *Mbo*I. The numbers are likely, however, to be a sufficient approximation for most purposes.

7. WORKING WITH LIMITING AMOUNTS OF DONOR DNA

The above procedures are most conveniently carried out with relatively large quantities of starting material (>100 μg of high molecular weight DNA). $5-10$ μg will be needed to determine the optimum partial digestion conditions. A considerable proportion of the partially digested material will fall outside of the desirable size range, and there may well be losses at various stages of the procedure, for example during dialysis or ethanol precipitation of salt gradient fractions. Moreover, it is not unknown for batches of restriction enzyme to carry additional, unspecified, nuclease activities that reduce further the efficiency with which recombinant DNA molecules are obtained. Finally, vector DNA preparations and DNA packaging extracts can both vary considerably from batch to batch. If starting material is not in any way limiting it will be possible to ignore some or all of these potential problems.

It is possible, however, to construct libraries starting from much smaller quantities of DNA, isolated perhaps from a non-renewable source such as a small tumour. Assuming maximum efficiency and no losses at every stage of the cloning procedure as little as 1 μg of size fractionated DNA can be sufficient for a library of greater than 10^6 recombinants (6). Unfortunately this efficiency is seldom obtained.

It has already been stressed that extreme care should be taken in the preparation of the genomic DNA. The method of DNA isolation described in Section 18.4 has been used to prepare as little as $10-20$ μg of DNA that can be cloned efficiently. Because it would be wasteful to use any of this DNA to determine the correct partial digestion conditions, the conditions should be determined on another, less important, but similarly prepared DNA. Then digest some of the sample under these conditions and analyse an aliquot (~0.5 μg) on a 0.3% agarose gel. If necessary, alter the conditions and digest the rest of the sample. Err on the cautious side, and stop the reaction by rapid freezing (not EDTA or phenol). It is always possible to add more enzyme if necessary.

Maximum efficiency might be obtained by treating the DNA fragments with phosphatase and using them directly for ligation to the appropriate vector DNA (Section 3.2), and by packaging the recombinant molecules under conditions that favour the recovery of small as well as large inserts (Section 4.2). If, however, the fragments are to be fractionated by centrifugation in sucrose or sodium chloride gradients this should be done in small tubes (e.g., Beckman SW51), and care should be taken during ethanol precipitation (Section 18.2). If it would be inordinately wasteful to analyse the fractions by electrophoresis, assay fractions from an appropriate region of the gradient for the efficiency with which their DNA can become incorporated into packagable DNA molecules.

Lastly, do not waste your precious DNA until you are convinced that each stage of the procedure is working at maximum efficiency. Pay particular attention to the phosphatase step if this is used (see Section 18.3) and use only the highest efficiency packaging extracts.

8. AMPLIFICATION

To a first approximation the growth characteristics of recombinant phage will be unaffected by sequence content and, assuming an appropriate choice of vector and host (Sections 15 and 16), all regions of the genome will be proportionally represented as viable phage. If some recombinants grow slightly faster or slower than the norm this may be reflected in plaque size, but the vast majority of recombinant DNA molecules will give rise to plaques, even the smallest of which can be detected by hybridisation methods.

A purpose-built library may be packaged, plated, screened, and discarded. Either the desired sequence will have been found, or it is not contained in the library. If necessary more DNA can be packaged and screening repeated. If high cloning efficiencies are being obtained routinely it might be argued that only purpose-built libraries are necessary. Cleaved donor and vector DNAs can be stored frozen and combined as required. On the other hand, it may be deemed prudent to produce a more permanent library at a time when everything is working and all the necessary ingredients are fresh and on hand.

Amplified libraries will maintain a fairly high titer for several years and can be produced easily (2), but are not without their drawbacks. They are useful, however, not just as a general resource from which many genes may be plucked over a period of time, but also when a library must be used repeatedly, as for example during 'chromosome walking' (Section 12). They are also convenient if a library is to be distributed to other workers. The major drawback of amplification is that differential growth of recombinants will be reflected in the sequence content of the amplified library, and the number of clones that has to be screened for a given p value will increase. Unfortunately this is not readily quantifiable, but should be borne in mind if the desired recombinants are not found. Proportional misrepresentation is exacerbated by the competitive growth of recombinants in liquid culture, or at very high plaque density. For this reason amplification is most commonly achieved by growth of the primary library at relatively low density as plaques on agar plates, such that individual plaques can still be resolved, and it is advisable to start with more than a 'complete' library. Amplification in this way will also reduce the likelihood that bacteria will be simultaneously infected with more than one recombinant phage, which could lead to genetic recombination between repetitive elements present on different inserts.

Amplification on a recombination-deficient (Rec$^-$) host should minimise both this problem and any tendency towards recombinational instability dependent on repetitive sequences within a cloned DNA fragment (Section 16.2). For detailed information on amplification of the recombinant derivatives of EMBL vectors see reference 6.

To minimise heterogeneity of plaque size and hence differential amplification, use fresh plating cells to which the phage have been pre-absorbed (see Section 13.2), and keep the plates absolutely horizontal until the top agar has set. After growth, overlay the agar surface with L-broth or phage buffer and leave for more than one hour at room temperature, or overnight at 4°C. Pipette the phage solution into a centrifuge tube, add a few drops of chloroform, and spin at low speed to remove bacterial debris. Titer the phage and store at 4°C. Conditions for the long-term storage of lambda are discussed in references 1 and 2. Amplified libraries may be concentrated by equilibrium centrifugation in CsCl gradients.

Re-amplification of such a library is inadvisable due to the potential for overgrowth by a sub-population of relatively vigorous phage.

9. SCREENING BY HYBRIDISATION

If all has been successful up to this point, the proud possessor of a genomic DNA library turns his or her mind to the problem of finding a particular DNA sequence of interest from among as many as 10^6 (or more) independent clones. This would be like looking for the proverbial needle in a haystack if no probe or selection system were available. For the purposes of this section we shall assume that screening is to be carried out by hybridisation methods and that a suitable probe, be it DNA or RNA, is available. Other methods of screening will be considered in Section 10.

In many cases the probe will consist of a DNA fragment that has been cloned using a plasmid as vector, and labelled by 'nick translation' or other means. It should be noted that a number of plasmid vectors in current use, notably cosmids and certain expression vectors, contain sequences derived from lambda DNA and which, if also present in the arms of recombinant DNA molecules, may obscure identification of the desired recombinant. In such cases it will be advisable to purify the probe fragment by agarose gel electrophoresis (2,13) either before or after labelling (see also Section 12).

Hybridisation is carried out, not on the plaques themselves, but non-destructively on filter replicas to which the plaques have been blotted (2,14). Nitrocellulose or nylon membrane filters can be used. A plaque of phage growth is a source, not only of amplified phage particles, but also of a considerable amount of free, unpackaged, phage DNA. The latter, even if somewhat degraded, remains a more or less adequate substrate for hybridisation. An agar surface in a Petri dish, or other receptacle (cafeteria trays have been used), on which an arbitrary proportion of the library has been plated out in an agarose top layer, is brought into contact with a sterile circle or sheet of filter, which absorbs moisture from the agarose. Both phage particles and free DNA will be absorbed by the filter and, if the agarose surface is not too wet, spreading will be minimal. When the filter is removed from contact with the agarose it carries an invisible replica of the pattern of plaques that are on the surface. Agarose instead of agar in the top layer discourages the top layer from being lifted off when the filter is removed, and may also reduce non-specific binding of the radio-active probe to the filter replica.

The replica, of which a number can be taken from one plate, must be treated in order that the phage DNA be denatured and fixed onto the filter. Nitrocellulose does not bind double-stranded DNA but has a very strong affinity for the single-stranded form. Brief treatment of replicas with strongly alkaline solutions causes phage particles to release their DNA, and causes all DNA on the filter to denature. The DNA remains single-stranded when a more neutral pH is restored by treatment with concentrated buffer solutions, and is irreversibly fixed to the filter by baking at 80°C. The filter is 'pre-hybridised' in a solution, the components of which saturate its non-specific DNA binding capacity, and can then be probed by hybridisation with a radio-actively labelled and single-stranded DNA (or RNA) species. Under conditions of temperature and salt concentration that favour the renaturation of homologous, or partially homologous nucleic acid species (discussed in reference 2), some proportion of the probe will become bound to regions of the filter that carry related DNA sequences. Authentic binding is resistant to washing in probe-free solutions and can be visualised by autoradiography (2,14).

It should be then a simple matter to match up the 'positive signals' with the corresponding plaques (or regions) on the agar plate. Authentic signals will appear in equivalent positions on both of a pair of duplicate filters, and thereby can be distinguished from non-specific binding of the probe which can appear in the form of small plaque-shaped spots.

Between 500 and 1000 small plaques on the surface of a 90 mm diameter agar plate can be independently resolved. In the case of an *E. coli-* or yeast-sized genomic DNA library, only $10^3 - 10^4$ independent recombinants may be required for 'complete' representation (p = 0.99). Two to twenty 90 mm plates, therefore, can carry the complete library. With luck a positive signal on the replica filter(s) will be found to correspond with a fairly well separated plaque, which can be picked independently of its neighbours.

Low density screening may even be feasible for a *Drosophila*-sized genome (3×10^4 recombinants for p = 0.99), but will never be economical for mammalian or plant genomes ($3 \times 10^5 - 3 \times 10^6$ recombinants necessary) because of the expense of filter membranes, and because of the inconvenience of handling so many plates. It is normal, therefore, to screen large libraries at a much higher density (10^4 or more per 90 mm plate equivalent). At this density, lysis is almost confluent across the surface of the plate and it is impossible to identify and pick one individual plaque. In this case a plug of agar is picked from the correct area of the plate, phage from the plug are resuspended in phage buffer, and dilutions are plated out such that 200 – 500 plaques are evenly distributed on the surface of an agar plate. Again filter replicas are taken, hybridised (keep the original probe for this: Section 18.10) and well-separated plaques that correspond in position to positive signals are picked.

Phage can diffuse across the surface of an agar plate, especially if it has been blotted with nitrocellulose filters. Therefore it is necessary at this stage, of both low and high density screening procedures, to repurify the picked plaque by a further round of dilution and plating. Alternatively, single plaques may be obtained by streaking (2). Pick at least two well-separated plaques, this time at random, and grow them on a small or large scale for DNA isolation and characterisation.

It is important to note that plating at near-confluence encourages the simultaneous infection of some bacteria (those in regions where plaques overlap) by more than one recombinant phage. Unless a Rec⁻ strain is used (Section 16.2), this can lead to genetic recombination between repetitive elements present on different DNA molecules. In turn this may lead to the generation of viable phage that carry sequences derived from two different regions of the genome. Recombination could be a particular problem for donor DNA carrying highly repetitive sequences such as the Alu elements (present every 10 kb or so in human DNA and hence present in a very high proportion of recombinant phage). If high density screening on Rec⁺ hosts is unavoidable, the potential for novel juxtapositions should be remembered when the recombinant DNA molecules come to be analysed, although the majority of the phage giving rise to a positive hybridisation signal should still be of the un-rearranged form.

In addition to plaque-shaped spots, other forms of background (non-specific) hybridisation may be encountered despite exhaustive washing. For example, blotchy or uniform distribution of radioactivity across the filter, or significant labelling of all plaques (only distinguishable as such in low density screening procedures). If an homologous probe is used to screen the library, this will seldom interfere with the recognition of positive signals. The use of heterologous probes, or complex mixtures of radioactive cDNA or messenger RNA, on the other hand, may be more problematic. Unfortunately, the very recombinant phage that one is trying to isolate is often the only true control for both signal to noise problems and, ultimately, the presence or absence of the desired sequence in the library. If no progress is being made with the screening, however, some form of control will be necessary to confirm that all steps of the procedure are being performed correctly.

(i)   *Control plaques:* If recombinant lambda derivatives homologous with the probe DNA are already available, these may be grown as plaques, at relatively low density, on a control plate. A filter replica is taken and used for hybridisation. Such control phage will be available, for example, during all but the initial stages of 'chromosome walking' (Section 12). It is also possible to create a control plaque (a 'macroplaque') by "tooth-picking" into the surface of a culture dish immediately after the library has been applied, and the top layer has set.

(ii)   *Control probe:* Filter replicas of the library are hybridised with DNA sequences that are known to be present on them. If the probe corresponds to a repeated sequence within the genome (such as a cloned ribosomal RNA gene), it will be necessary only to screen a small proportion of the library.

A specific protocol for screening by hybridisation is given in Section 18.10.

10. ALTERNATIVE SCREENING AND SELECTION PROCEDURES

Although screening by plaque-hybridisation, using a labelled DNA or RNA probe, is the most generally useful method, it is not necessarily the only one. An alternative selection system has been devised by Brian Seed and his colleagues (2). A genomic DNA library is prepared as described above using a vector that carries two amber mutations within essential lambda genes. The library is allowed to infect a recombination-proficient (Rec$^+$) *E. coli* strain that harbours the 'probe' fragment present in a plasmid DNA molecule. In addition to the probe fragment the plasmid encodes a suppressor (a mutant tRNA) capable of suppressing amber mutations. Homology between the probe and a particular recombinant DNA molecule permits the plasmid to recombine with, and hence integrate into, the phage genome. Phages carrying the plasmid are then selected from the lysate directly by their ability to form plaques on a suppressor-free host. For this method to be generally useful the integrated plasmid must be shed (by an excisive recombination event) before the cloned DNA is analysed or further manipulated. Partial (i.e., interspersed) homology between the probe and the donor DNA fragment is, therefore, a potential source of difficulty. This method also suffers the drawback that the library must be propagated under Rec$^+$ conditions (Section 16.2). A derivative of EMBL3 (EMBL3a) containing two amber mutations is available but for unknown reasons the efficiencies of selection (6) have not equalled those reported by Seed (2) for the vector Charon 4A.

Screening procedures may also be based on the expression of cloned DNA sequences. This may depend either on the complementation of defects of the host bacterium, or on the immunological detection of polypeptides (discussed by Murray in reference 1). The replacement vectors shown in *Figure 8* include good promoters from which cloned DNA sequences may be transcribed, but they make no special provision for the processing and translation of the mRNA. Such methods are more likely, therefore, to be possible for prokaryotic genes, although some successes have been reported with genes from lower eukaryotes.

11. ANALYSIS OF RECOMBINANT DNA MOLECULES

Having purified one or more recombinant phages with the desired hybridisation or biological properties, it will usually be necessary to characterise their genomes in order, for example, to define their exact relationship to the probe DNA and to each other, and to develop subcloning strategies that will enable further characterisation and/or manipulation of the sequence(s) of interest. Methods relevant to the large scale preparation of lambda DNA molecules, and to their physical characterisation (restriction enzyme analysis, agarose gel electrophoresis, Southern blotting, heteroduplex analysis etc.) are described in references 1,2,13,15,16 and 17. We shall mention only a few points of special relevance to the analysis of recombinants obtained as described in this article.

(i)   A preliminary step in the characterisation of recombinant DNA molecules will often be the construction of a 'restriction map' showing the positions of targets for enzymes that cut relatively infrequently within the inserted DNA (invariably enzymes that recognise a hexanucleotide DNA sequence). This task will be made considerably easier if an overlapping set of clones is available, as regions of overlap between any pair of inserts will give clues to the relative positions of specific restriction fragments within them (*Figure 5*). The use of an enzyme that cleaves within the retained section of the polylinker (e.g., *Sal*I in the case of EMBL3:*Figure 4*) will enable fragments from the extremities of the inserts to be recognised. These, unlike fragments entirely contained within the inserts, will vary in size from molecule to molecule (because of their 'random' ends). It is important to remember that if the library was screened at near-confluence in a Rec$^+$ host, novel juxtapositions of DNA within the inserts can occur (Section 9).

(ii)   Oligonucleotide sequences complementary to each of the protruding *cohesive ends* of lambda DNA are available, and may be used to label either end of the genome. This method has been used to expedite restriction mapping via a labelled 'ladder' of partial digestion products (18) that have been separated by electrophoresis. The ladder may be read from either end of the genome. This type of analysis will, of course, be complicated by the presence of targets within the vector arms, or by the presence of many closely spaced targets within the inserted DNA. A vector containing the *cos* sequences close to the cloning site would allow considerably greater resolution.

(iii)   The central fragment of the EMBL vectors shares homology with the right arm of the vector DNA (6). This can give confusing results if recombinant DNA molecules are analysed by heteroduplexing with the vector DNA (H. Delius, personal communication). For this reason a recombinant DNA molecule will often be preferable to vector DNA when a control is needed for heteroduplex analysis.

12. 'CHROMOSOME WALKING'

Libraries made as described above are ideally suited to chromosome walking (19,20). There are two reasons for this. Firstly, they consist of an overlapping and equally representative set of genomic DNA fragments so, at least in theory, any point on a chromosome can be reached by 'walking' from any other point on the same chromosome. Secondly, restriction sites that flank the *Bam*HI sites of the vector are retained in recombinant phage DNA molecules. For example, the *Bam*HI sites of EMBL3 are flanked immediately on their outsides by *Sal*I sites (*Figure 4*). *Sal*I thus releases cleanly the inserted DNA from its association

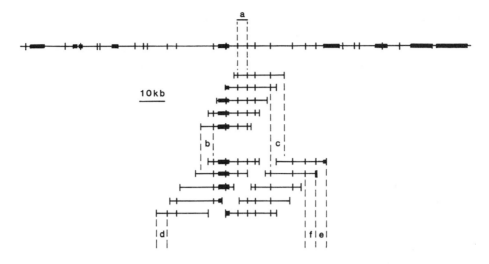

Figure 5. Chromosome walking. In this hypothetical case, a cloned *Eco*RI fragment containing no repeated DNA sequences (fragment **a**) is used to initiate a walk in both directions along the chromosome in which the fragment normally resides. A vertical line indicates the position of a *Sal*I site (within DNA molecules) or a site within the polylinker that allows clean separation of inserted and vector DNA sequences (at the ends of the overlapping fragments). Each black box represents a member of a different family of repeated DNA sequences, the other members of which lie elsewhere on the same or different chromosomes. Probe **a** is used to identify a set of overlapping inserts in recombinant phage DNA molecules. Only their inserts are shown. Fragments **b** and **c**, which lie at the extremities of the newly cloned regions are used to identify two further overlapping sets of inserts, and so on with fragments **d** and **f**. Fragment **e** cannot be used as it contains repeated DNA sequences and will cause the walk to 'branch'. A very long section of repeated DNA (at the extreme right) might only be overcome by using a cosmid recombinant.

with the lambda arms. Sometimes there will be further sites for *Sal*I within the inserted DNA. In any case it will be possible to identify specific fragments (of the form *Sal*I-*Sal*I, or *Sal*I-X where X denotes another enzyme that cuts within the insert) that can be used as probes to pull related clones out of the library. If the probe fragments contain sequences from the extremities of the insert this will initiate a walk in both directions away from the region of the chromosome represented in the original recombinant (*Figure 5*). (N.B. *Bam*HI is not generally useful for the separation of arms from inserts since only one in four of the *Bam*HI/*Sau*3A hybrid recognition sites can be cleaved by this enzyme.)

For each of the lambda vectors shown in *Figure 8*, appropriate cleavage sites within the polylinker can be chosen that will cleanly separate vector and insert vector DNAs. DNA fragments chosen as probes can be purified away from the lambda arms and other unwanted fragments by molecular cloning using as vector a plasmid that has no homology with lambda DNA. Alternatively, the fragments can be extracted or eluted from an agarose gel by one of the methods described in references 2 and 13. Unless an unfortunate distribution of restriction enzyme sites makes it difficult to resolve the probe fragment from an unwanted fragment(s), the latter method is preferable because it is so much faster.

Specific activities sufficient for screening lambda libraries (as little as 2×10^5 to 1×10^6 d.p.m./μg: see reference 19) can be obtained by using the large fragment of *E. coli* DNA polymerase I to 'repair' the cohesive termini produced by cleavage with a restriction enzyme (end-labelling). One to four ^{32}P-deoxynucleotides can be incorporated at each terminus. If the cleaved recombinant DNA is end-labelled *before* electrophoretic separation, fragments isolated from the gel may be used as probes directly. This also avoids potential problems due to substances present in agarose and which may inhibit enzymatic (e.g., labelling) reactions.

In most cases the small section of the linker retained in the probe fragment will hybridise insignificantly to homologous sequences present in every recombinant phage genome, and thus contribute little, if any, background. It should be noted, however, that the *Eco*RI site in the linkers of the replacement vectors Charon 34 and 35 is 55 bp from the cloning (*Bam*HI) site (see *Figure 8*). This may be sufficient to produce a noticeable, and possibly inconvenient, background.

The general approach to genome walking is as follows:

(i) A library is screened with a particular unique sequence probe DNA (probe a, *Figure 5*). Several plaques that correspond in position to positive autoradiographic signals are picked, purified, and their phage are grown up on a small scale. Phage DNAs are isolated and are analysed by restriction enzyme cleavage and agarose gel electrophoresis. As the probe was a unique DNA sequence, all the DNAs should show some degree of relatedness with each other in the region homologous with the probe. This can be checked by Southern-blot (2,13) analysis of the gel, using the same probe DNA. It should be possible to recognise an overlapping set of insert fragments and to identify specific fragments in certain inserts (e.g., fragments b and c in *Figure 5*) that lie at the extremities of the cloned region. Analysis may be simpler if the lambda arms are allowed to anneal before electrophoresis.

(ii) Because of the flanking sites retained in the recombinant DNA molecules, b and c can be separated cleanly from the lambda arms. Phages containing these fragments are grown up, their DNA is purified, and b and c are eluted from a gel and labelled. Using them to probe the library produces two new sets of overlapping clones which must be analysed in the same way as the first. Terminal fragments d and e are identified and isolated for use as probes for the identification of further overlapping clones (*Figure 5*).

(iii) This simple, if somewhat time consuming, procedure will be complicated by the presence of repetitious DNA sequences such as bacterial transposons, copia-like elements or Alu sequences. In the hypothetical case of *Figure 5* several repetitive sequences occur (they are indicated by shaded blocks). Fortuitously, the insert from which probe b was obtained extended beyond one of these elements. Probe e, however, contains repeated DNA and will identify inserts from regions of the genome where other members of this repeated gene family occur. This can be avoided by probing instead with fragment f, in which case it should be possible to jump over the repeated element into a region of unique sequence DNA. This approach is limited however, as long regions of repeated DNA (>20 kb) will not be overcome, except by recourse to a cosmid library. Luckily repeated DNA elements of this length do not seem to be very common. The more abundant repeated DNA elements can often be recognised in recombinant phages by plaque hybridisation, or in particular phage DNA fragments by Southern blot analysis, in which total genomic DNA is labelled and used as the probe (19).

Some workers have overcome the problem of repeated sequences by taking advantage of their heterogeneity in different natural isolates of *Drosophila melanogaster* (20). A walk initiated within one strain, upon encountering a repeated DNA sequence, could sometimes be continued in a second strain which had no repeated DNA at that position on the chromosome, although it carried related sequences elsewhere in its genome. Polymorphism, both of restriction enzyme sites and of inserted sequence elements, may also occur within a single genomic DNA library if the donor DNA was not prepared from an inbred line.

Inevitably, the ease of walking within any given chromosomal domain will be inversely related to the frequency of repetitive DNA sequences found there. Walking is likely to be especially tedious in higher plants and animals. In the human genome, for example, the Alu sequences are present on average once every 10 kb or so.

(iv) If such a walk were to be made within a genome as small as *E. coli* ($< 10^3$ plaques for p = 0.99), it might be accomplished most quickly by repeatedly screening the same ordered array of plaques rather than plating out the library each time a new round of screening had to be carried out. An ordered array of 1000 independent plaques can be carefully created on the surface of a 20 x 20 cm square culture dish by toothpicking, and several identical membrane replicas can be taken. One replica is used for the initial screening (probe a). Positive signals are well separated from their neighbours and can be picked and grown up immediately without further purification of the phage, especially if a non-replicated master copy was toothpicked at the same time as the first. Non-specific background spots and other hybridisation artefacts are much less of a problem when working with an ordered array, so only one replica is needed per probe (nitrocellulose filters are expensive). Second and third replicas are screened with probes b and c. Each replica membrane may be probed several times. If the culture dish is maintained sealed and inverted (to prevent condensation collecting on the surface of the agar) at 4°C, phage remain viable for more than two months.

Not all of the plaques identified by probes b and c need be analysed at the DNA level however, since some of them will correspond to plaques already identified by probe a. This introduces a certain amount of directional bias to the walk. Moreover, the array may be screened simultaneously from several start points.

To maintain such an array of plaques on a longer term basis they may be either repicked every two months or so from the non-replicated master plate, or a pronged replicating block can be devised, and used to transfer plaques from dish to dish. Although it is tedious to create an array, the ease of replication afforded by a replicating block may be sufficient to justify organising the clones of a yeast-, or even a *Drosophila*-, sized, library into a matrix that may be replicated. On a smaller scale, entire bacterial genomes could be mapped, simultaneously from a number of sites, by this method. The continued replication and/or long term storage (as plaques) of a library can lead to problems due to the growth of lambda-resistant *E. coli* within plaques, and of a colourful variety of other bacteria and fungi elsewhere on the plates. This can be avoided to some extent by exposing the surface of the master-plate to chloroform vapour before using the replicating block, or before storage.

13. BACTERIOPHAGE LAMBDA

13.1 Relevant Biology

The linear genome of the lambda phage is packaged in an icosahedral head from which projects a tail ending in a tail fibre, the product of the lambda gene *J* (the gene that determines host-range). Lambda phage normally remain infective for years if kept at 4°C in phage buffer or L-broth containing $1 - 10$ mM $MgSO_4$, and in detergent-free glassware. The phage particles adsorb by the tip of the tail fibre to receptor sites in the outer membrane of the cell. These receptors, which are encoded by the *E. coli* gene *lamB*, are also required for the uptake of maltose, and their production is stimulated by growth in medium containing maltose as the carbon source, but lacking glucose so as to avoid catabolite repression. Adsorption is temperature-independent and is facilitated by magnesium ions. In contrast introduction of the phage DNA into the cell, which is dependent on a component of the inner membrane, occurs rapidly at 37°C and only slowly at low temperatures.

After entering the cell the lambda genome circularises via its *cohesive ends* and is converted into a covalently closed circular molecule, the substrate for both replication (*Figure 7*) and transcription. Lambda, being a *temperate* phage, may follow either the productive (lytic) or temperate (lysogenic) pathway. In either case, transcription is initated from two 'early' promoters, p_L and p_R (see *Figure 6*) to provide functions essential for DNA replication, genetic recombination, establishment of lysogeny and the transcriptional activation of the late genes. In a productive (lytic) infection transition from early to late transcription is achieved, virion proteins are made, the replicated genomes are packaged, and lysis of the cell releases about 100 infective particles. In the temperate response on the other hand, most phage functions become repressed, either directly or indirectly, by lambda repressor molecules bound to the operator regions associated with p_L and p_R. If repression occurs in time to prevent activation of the late genes, lysis is avoided and lysogeny may ensue. Stable, as opposed to abortive, lysogeny also requires integration of the lambda genome into the host chromosome by site-specific recombination in order that the phage genome (the *prophage*) is replicated as part of the *E. coli* chromosome. However, all the lambda vectors described

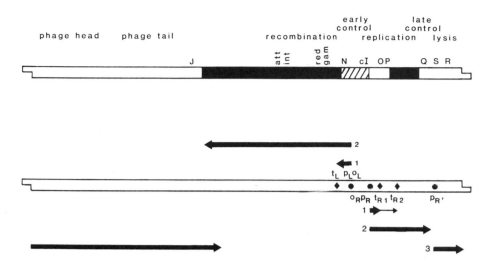

Figure 6. **(a)** Genes which encode related functions are clustered in the lambda genome. The black regions are inessential for propagation of lambda, and have been deleted during construction of the vectors described here (*Figure 8*). The hatched box is also inessential if the termination site t_{R2} (see below) is absent. The *cI* gene, encoding the lambda repressor, is deleted from the vectors described in this chapter, so both the vectors and their recombinant derivatives are obliged to embark on a lytic growth pathway (*Figure 7*). **(b)** Transcription of the genome proceeds initially from the promoters, p_L and p_R. Early during infection transcripts initiated at p_L and p_R (1) terminate at t_L and t_{R1} respectively (some rightward transcripts escape termination at t_{R1} and reach t_{R2}). This allows synthesis of the *N* gene product, in the presence of which transcription may proceed through t_L and t_{R1} into adjacent genes (2). This in turn allows synthesis of the *Q* gene product in the presence of which 'late' transcription (3), initiated from $p_{R'}$, continues through the lysis genes, past the *cos* site (of the circular genome: *Figure 7*), through the head and tail genes, and past the tail fibre gene *J*. Thus gene *N* (and its product) are normally essential for late gene activation. If t_{R2} is deleted, however, *N*-independent 'leakage' of transcription through t_{R1} can provide Q functions. Most lambda vectors, including those described here, make extra space by including a deletion (*nin5*) that removes t_{R2}. This deletion makes *N* and the p_L promoter dispensable for phage growth, however, the vectors described in *Figure 8*, and their recombinant derivatives, are N^+ so that they do not have to rely on transcriptional leakage and, therefore, grow relatively vigorously. ($\bullet$) major promoters; ($\blacklozenge$) major termination signals.

in this chapter lack both the repressor gene and the site-specific integration system, and therefore are obliged to embark on lytic growth.

In lytic growth, activation of the late genes of lambda is normally associated with a change in the predominant mode of DNA replication (*Figure 7*). During the first 15 min after infection, bi-directional replication of DNA yields monomeric circular daughter molecules (theta-mode replication). At later times the predominant products of replication are linear concatenated molecules presumed to originate by a rolling-circle mechanism. Concatenated genomes are the substrate for both *in vivo* and *in vitro* packaging, processes which involve the cutting of two appropriately spaced *cos* sequences and packaging of the resultant linear (mature) lambda genome. The transition to the rolling circle mechanism is impeded by the action of exonuclease V, the product of the *E. coli recB* and *recC* genes. In *recBC*⁺ cells concatenated DNA can still be produced, however, if the phage possess a functional *gam* gene. The product of *gam*, which is expressed soon after infection, binds to exonuclease V thereby inactivating its nucleolytic activi-

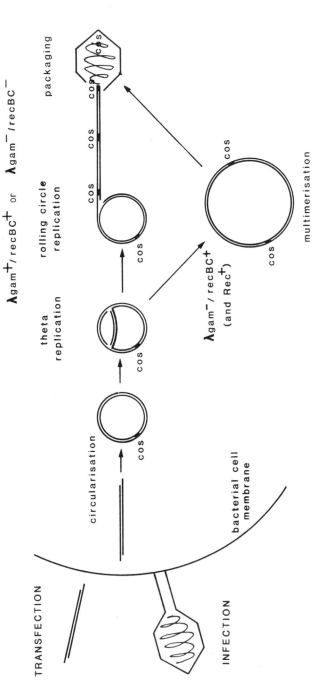

Figure 7. The lytic pathway of bacteriophage lambda. Phage DNA molecules may be introduced into a bacterial cell either in the form of naked DNA (transfection) or by infection with a phage particle. Inside the cell the linear DNA circularises via its *cohesive ends*, and a covalently closed circular DNA molecule is formed by the action of *E. coli* DNA ligase at the now double-stranded *cos* sites. For the first 15 min or so after infection, replication proceeds bi-directionally (theta-replication) to generate a number of monomeric circular copies of the phage DNA. The transition to rolling circle replication from the circular templates is blocked in *recBC⁻ E. coli* unless the phage produces a functional *gam* gene. A *gam* gene is unnecessary for the transition if the *recBC* nuclease is defective. The product of rolling circle replication is a linear concatenated DNA molecule, the usual substrate for packaging *in vivo*. Monomeric circles cannot be packaged. During processing of the concatenated DNA, the packaging machinery recognises two suitably spaced *cos* sites (40–52 kb apart) at which are made staggered breaks, thereby creating a mature linear phage genome, bounded by *cohesive ends*, and assembled into an infective phage particle. Phage that lack a *gam* gene can be propagated in *recBC⁺* cells, albeit less efficiently, if a functional generalised recombination pathway is available. Recombination between monomeric circular forms generates multimeric circular molecules, which *are* substrates for packaging.

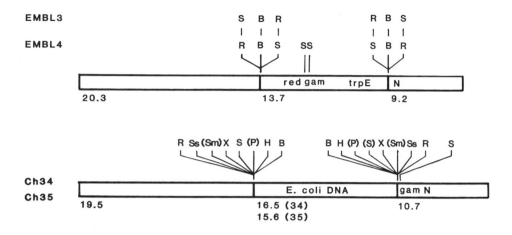

Figure 8. Replacement vectors discussed in the text. Sizes of the arms and stuffer fragments (in kb) are indicated. **(a)** EMBL3 and EMBL4 were derived by a number of steps from the vector 1059 (29). 1059 has no polylinkers and carries ColE1 plasmid DNA sequences in the central fragment. Background due to parental sequence phage is thus a considerable problem when libraries in 1059 are screened with plasmid probes. To alleviate this problem the central fragments of EMBL3 and EMBL4 contain the same lambda sequences as does the central fragment of 1059 (some of which are duplicated in the right arm), fused to the *trpE* gene of *E. coli*K12 instead of to ColE1. **(b)** Charon 34 and 35 (8) are identical except for their central fragments. Their polylinkers (60 bp) contain a number of sites for restriction enzyme cleavage, not all of which are absent from the vector arms. Insertion at the *Eco*RI, *Sst*I, *Xba*I, *Hind*III, and *Bam*HI sites, which do not occur in the arms, produces recombinant DNA molecules whose inserts may be excised by cleavage within the polylinker. *Sal*I may also be used as the 'cloning' enzyme, in which case a *Sal*I site in the right arm is used. Although they cannot be used for cloning, the *Sma*I and *Pst*I sites in the linkers can both be used to excise inserts. These four vectors are the sophisticated accumulation of many years genetic and biochemical manipulation of the lambda genome. For a complete knowledge of their structure the reader must refer to original sources. Without this knowledge it is not possible to predict a restriction map with any degree of certainty. Fortunately the predicted (and in some cases tested) maps for many commonly used restriction enzymes may be found in references 6 and 8. All the vectors in *Figure 8* are cI⁻ (△KH54), and all carry the *nin*5 deletion that removes the termination site t_{R2} (*Figure 6*). **(c)** The abbreviations for restriction sites are B, *Bam*HI; H, *Hind*III, P, *Pst*I; R, *Eco*RI; S, *Sal*I; Sm, *Sma*I; Ss, *Sst*I; X,*Xba*I; EMBL3 and EMBL4 are being made available by the American Type Culture Collection (see footnote [c] to *Table 2*).

ty (but not necessarily its ability to participate in genetic recombination).

All of the vectors suggested here are *gam*⁺. It should be noted, however, that in order to provide more cloning capacity, and to allow for a genetic selection against non-recombinant phage during propagation of the library (Section 14), recombinant derivatives of EMBL vectors sacrifice the *gam* gene carried by the parent vectors (*Figure 8*). The *gam*⁻ recombinants *can* be propagated in *recBC*⁺ hosts, however, as long as they can undergo molecular recombination to produce di- or multi-meric (and therefore concatenated) circular DNA molecules (*Figure 7*). This cannot be achieved by phage-encoded recombination enzymes (Red) as the *red* genes, which normally reside immediately adjacent to *gam* in the lambda genome (*Figure 6*), are also lost with the central fragment. Hence the *red*⁻ *gam*⁻ recombinants rely, perforce, on host recombination (Rec) functions.

Paradoxically, as long as the host is *recA*⁺ as well as *recBC*⁺, recombination will proceed by the RecBC pathway in which exonuclease V participates (Section

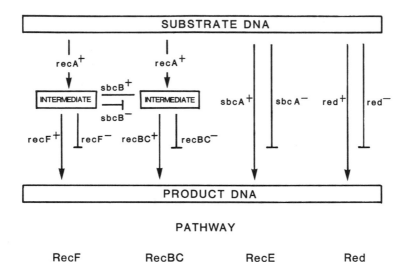

Figure 9. Recombination pathways relevant to our discussion (model and figure adapted from reference 21). The RecBC pathway is the predominant (99%) recombination pathway in wild-type *E. coli*K12. *recB⁻*, *recC⁻* or *recB⁻recC⁻* (*recBC⁻*) strains are defective in this pathway. Generation of an *sbcB⁻* mutation in any of these strains appears to prevent shunting from the RecF to the RecBC pathway, enabling efficient recombination (~50% of the wild-type level) to be catalysed by the *recF* product. *recA⁻* strains, which are defective in both pathways, are the least able to catalyse recombination ($10^{-3} - 10^{-6}$ of wild-type). The RecE pathway is an independent pathway to recombination. Access to this pathway is gained by an *sbcA⁻* mutation. In terms of its biochemistry and genetics, the RecE pathway resembles, in many respects, the Red recombination pathway encoded by wild-type lambda. This is because the *recE* gene is carried on a defective lambdoid prophage present in the chromosome of most *E. coli*K12 strains. Normally the *recE* gene is repressed (as are the *red* genes of a lambda prophage). *sbcA⁻* mutations, which map close to *recE*, in some way awaken the *recE* gene from repression. The *recBC⁺*, *sbcB⁺*, and *recE⁺* (= *sbcA⁻*) products are respectively, exonucleases V, I, and VIII.

16.2 and *Figure 9*). Only phage DNA molecules which possess one or more copies of an octanucleotide sequence known as Chi are an efficient substrate for the RecBC pathway (1). Wild-type lambda has no Chi sequences (Red is Chi-independent). In order that their recombinant derivatives may be propagated in *recBC⁺* hosts (such as P2 lysogens: Section 14), EMBL vectors have been endowed with a Chi sequence in the vector arms. Certain replacement vectors, such as Charon 4, Charon 4a and Charon 30 (1,2,4,5), are both *gam⁻* and *Chi⁻* in the absence of the central fragment, and hence recombinants that carry a Chi sequence in the donor fragment are at a considerable growth advantage over those that do not. These vectors can only be recommended for the construction of representative libraries if the libraries are to be (or have been) recovered, or amplified, on *recBC⁻* hosts. Charon 34 and Charon 35 (*Figure 8*) generate *gam⁺* (and *red⁻*) recombinants that have no need of a Chi site.

13.2 Factors Influencing Plaque Size

Efficient adsorption of phage lambda to *E. coli* requires the *lamB*- encoded receptor sites in the outer membrane of the cell, and is facilitated by magnesium ions. Phage will adsorb to dead cells and debris, however, so cultures that have been in stationary phase for some time, and thus contain a high proportion of dead cells, are better avoided.

To ensure optimum adsorption, start with a single colony of the appropriate bacterium and grow the bacteria in L-broth to a concentration of approximately 2.5×10^8 cells per ml (an OD_{650} of ~0.5), in the presence of 0.4% maltose but in the absence of glucose. Concentrate the bacteria by low-speed centrifugation, and resuspension in one half to one tenth the volume of 10 mM $MgSO_4$. Aerate by shaking at 37°C for 1 h. Starved cultures containing Mg^{2+} can be stored at 4°C for several days without much loss in viability unless the cells are Rec⁻.

Incubation of phage with the cells at room temperature for 15 min promotes efficient (>90%) adsorption and, when the temperature is raised to 37°C, entry of the DNA occurs rapidly. Pre-adsorption is important because it synchronises infection and thereby reduces heterogeneity of plaque size.

Invariably a phage solution must be diluted, firstly so that it may be titered, and secondly so that a predetermined number of phage can be added to the bacteria (even a mere plaque will contain $10^5 - 10^7$ phage particles). Phage may be diluted in either nutrient broth or in phage buffer (see Section 18.1 for recipes), although broth may provide nourishment for invading microorganisms on long-term storage.

To obtain plaques, $0.1 - 0.2$ ml of cell suspension (at ~10^9 cells/ml) to which an appropriate number of phage have been adsorbed are mixed with $2 - 3$ ml of molten 0.65% nutrient agar or agarose maintained at 50°C. The mixture is spread immediately, with a gentle swirling motion, as a top layer on the surface of a 90 mm diameter nutrient agar ($1 - 1.5$%) plate, and allowed to set. If different sized culture dishes are used, the amounts should be adjusted accordingly. Both the bottom and top layers should be poured on a horizontal surface. Otherwise both plaque size and plaque density will vary across the surface of the plate. Incubate the plates upside-down to prevent condensation collecting on the agar surface. Include a control plate to which no phage have been added as a check on contamination (with exogenous phage *or* bacteria), and on bacterial growth. After growth at 37°C for eight hours to overnight, plaques ($0.1 - 2$ mm diameter) should be apparent as clear spots against an opaque background of unlysed bacteria. Agarose in the top layer is necessary only if filter replicas are to be taken (Section 9), but is slightly more difficult to pour because it sets more quickly than agar at the same temperature. For this reason, plates taken from the cold should be allowed to warm up, at least to room temperature, before pouring the top layer.

Plates that carry an excessive amount of surface moisture are not a good substrate for the top layer as plaques may be encouraged to spread or even 'run' across the surface of the plate, and the top layer may be lifted more easily from the plates when filter replicas are taken. To avoid these problems the bottom layer may be dried, with the lid off and if possible in a sterile environment, until a slight puckering of the surface is apparent. Alternatively pour the plates several days in advance.

In general, any factor that increases the time taken for the growth of a confluent lawn of bacteria will increase the average plaque size, since plaques stop increasing in size once the lawn is fully grown (i.e., once the cells have reached stationary phase). Plaques will, therefore, be larger on relatively less rich plates, or if fewer cells are added. A second generalisation is that larger plaques are favoured by a moist environment e.g., fresh plates, lower agar concentration in the bottom layer, more top-layer, or in a humid overcrowded incubator rather than a 37°C room.

To restrain plaque size, as may be desirable when screening a library at high plaque-density, choose a hard and rich medium (e.g., L-broth agar) for the bottom layer, and L-broth agarose (0.65%) for the top layer. On the other hand, red^- gam^- phage (Section 14) produce such small plaques, especially on hosts lysogenic for P2, that the problem may be to increase plaque size. It is not unusual for a novice to fail to recognise the minute plaques of a red^- gam^- phage if excess cells or a rich, dry medium has been used. In such cases use a less rich and softer medium (e.g., BBL-agar) for the bottom layer, and a BBL trypticase-based top layer. $recBC^-$ strains, which are slow-growing and allow rolling-circle replication (see Section 16.2), will also allow red^- gam^- phage to form large plaques (but use of such a strain would preclude the use of a P2 lysogen).

It is worth remembering that

(i)   Even the smallest plaques are likely to contain sufficient phage DNA to be detectable by hybridisation methods.

(ii)   Although a $recBC^-$ strain may foster large plaques, it will not necessarily yield high titer lysates because of its subnormal vigour. Inevitably, when the predominant genetic recombination system is blocked, the phage yield is reduced. When growing red^- gam^- phage for the purpose of DNA isolation, the highest titer lysates, both on plates and in liquid culture, are likely to be achieved by growth in the Rec^+ *E. coli* derivatives $recBC^-$ $sbcA^-$ or $recBC^-$ $sbcB^-$ (Section 16.2).

14. SELECTION AGAINST PARENTAL SEQUENCE PHAGE

gam^+ phages are unable to form plaques on *E. coli* lysogenic for phage P2 (see reference 1). Although incompletely understood, this phenomenon can be turned to advantage by including a functional *gam* gene in the central fragment of an otherwise gam^- replacement vector. EMBL3 and EMBL4 (Section 15) are two such vectors for which it is, therefore, unnecessary to remove physically (or otherwise inactivate) the central fragment before the ligation of vector arms to donor fragments. Parental sequence genomes (gam^+) will be among the products of the ligation reaction, but they can be effectively discriminated against by propagating the library on an *E. coli* (P2) strain. Only gam^-, and hence recombinant, phages *should* form plaques on such a strain.

Sensitivity to *P2 I*nterference (the Spi$^+$ phenotype) appears to be due, at least in part, to inactivation of the *E. coli recBC* nuclease by the *gam* gene product of lambda (Section 13.1). Phage P2 cannot lysogenise a $recBC^-$ strain and infection of a P2 lysogen by gam^+, but not gam^-, lambda leads to cessation of both protein synthesis and DNA replication. A gam^- lambda, providing it is also Chi$^+$, will grow in hosts lysogenic for P2, but larger plaques are obtained if the phage is also *red⁻* (recombination deficient). Such a *red⁻ gam⁻* Chi⁺ phage is said to have a Spi⁻ phenotype, since it is insensitive to interference by the P2 prophage. The relevance of *red* to this phenotype is not known. Nevertheless, the Spi⁻ phenotype is commonly used to select *red⁻ gam⁻* derivatives of a Chi⁺ phage and is the basis for the selection of recombinant derivatives of Chi⁺ replacement vectors (e.g., EMBL3 see *Figure 8*) in which the central dispensable fragment includes both the *red* and *gam* genes of lambda.

15. CHOOSING A REPLACEMENT VECTOR

We have limited our discussion to vectors that allow donor DNA cleaved with *Sau*3A to be inserted at *Bam*HI sites, and at the same time permit the separation by enzymatic cleavage of cloned sequences free of vector DNA (*Figure 8*). Alternatively, but usually less efficiently, any of the same vectors cut with *Eco*RI could serve to clone a random population of mechanically sheared DNA fragments to which had been added *Eco*RI linkers. Furthermore, all the vectors shown in *Figure 8* can be used for cloning fragments generated by cleavage of donor DNA with other enzymes (e.g., *Eco*RI-, *Eco*RI*-, *Bam*HI-, *Bgl*II-, *Bcl*I-, *Xho*II-, *Sal*I-, *Xho*I-, and a subset of *Ava*I- fragments in the case of EMBL3) although in most cases this will not give rise to a representative library.

EMBL3 and EMBL4 (6) are identical except for the orientation of their polylinkers. Both will accept donor fragments in the 9 − 22 kb range and, because their recombinant derivatives may be selected genetically in P2 lysogens by virtue of their Spi⁻ phenotype (Section 14), there is no need to remove the central fragment. A simple biochemical method for inactivating the central fragment of EMBL3 (Section 3.4) has been sufficiently effective that up to 98% of packageable DNA molecules are recombinant. This method, or physical removal of the central fragment, obviates the need for a P2 lysogen which is of necessity *recA⁺* and *recBC⁺*, and allows the *red⁻ gam⁻* recombinant derivatives to be selectively recovered under Rec⁻ conditions (Section 16.2). Both vectors have two *Sal*I sites within the central fragment. Cleavage of EMBL4 with *Sal*I, in addition to *Bam*HI, even if it does not enzymatically isolate the stuffer fragment (discussed in Section 3.4), would aid resolution of the arms during their physical separation. Detailed restriction maps of EMBL3 and EMBL4 may be found in the Appendix to reference 1.

Charon 34 and Charon 35 (8) differ from each other only in their central fragments, and will accept donor fragments in the 9 − 20 kb range. These vectors have a more extensive range of restriction targets within their polylinkers than do EMBL3 and EMBL4, and their recombinant derivatives, being *gam⁺*, are able to grow in any Rec⁻ host. However, both vectors rely on the physical separation of lambda arms from the central fragment before ligation to donor fragments. Charon 35 has two *Bam*HI sites within the stuffer which may aid separation of its arms. Detailed maps of both vectors are given in reference 8. Both Charon 34 and Charon 35 have an additional *Sal*I site within the right arm (*Figure 8*). Nevertheless they can be used for cloning *Sal*I fragments as no essential DNA is deleted.

A further consideration applies to the choice of vectors for the propagation of vertebrate and plant DNA, which are relatively deficient in the dinucleotide 5'-CG-3'. CG is at the heart of the *Sal*I recognition sequence (GTCGAC), and for this reason *Sal*I sites are present less often than might be expected in these DNAs. If the DNA is cloned using a vector in which *Sal*I sites flank the *Bam*HI site (such as EMBL3), then it will be possible, by cleaving the recombinant DNA molecule with *Sal*I, to recover the entire insert on the minimum possible number of fragments (with luck only one). This would facilitate subcloning of the entire insert using the *Sal*I site of a plasmid vector.

16. CHOOSING A BACTERIAL HOST

Examples of *E. coli* strains with the genetic characteristics discussed in this section are listed in *Table 2*.

16.1 **Host Restriction and Modification**

All strains liable to be used as hosts will be derivatives of *E. coli* K12, which normally produces a restriction enzyme (*Eco*K) that will cleave DNA should it contain unmodified *Eco*K recognition sequences. *E. coli* K12, of course, also produces a modification methylase that protects its own DNA from cleavage. Unmodified *phage* DNA will, however, be subject to cleavage, as will the inserts of recombinant phage DNA molecules even if the vector has been grown in a modifying strain. This can be avoided by propagating recombinant phage in a restriction deficient (r_k^-) host. *hsdR*$^-$ strains are defective in restriction but not in modification ($r_k^- m_k^+$) while *hsdS*$^-$ strains are deficient in both functions ($r_k^- m_k^-$). The r_k^- phenotype can be recognised easily by the inability of the cells to distinguish between modified and unmodified lambda derivatives. Unmodified lambda phage form plaques approximately three orders of magnitude more efficiently on r_k^- than on r_k^+ hosts.

16.2 **Host Genetic Recombination Systems**

E. coli K12 has the ability to undergo more or less efficient homologous recombination by a number of different pathways (*Figure 9*). The pathways dependent upon the *recBC*$^+$ and *recF*$^+$ products are also dependent upon the wild-type product of the *recA* gene, whereas the RecE pathway is not. Only the RecBC$^+$ pathway is stimulated by Chi sequences (Section 13.1).

The recombination characteristics of *E. coli* strains are particularly relevant to the choice of hosts for amplification or screening of a library (see also discussions in Sections 8 and 9) and for the subsequent large-scale culture of purified recombinant phages (for DNA preparation). For example, homologous recombination between misaligned repeated sequences in lambda has been shown to lead to duplication and deletion derivatives, although some constraint on this process will be imposed by the size of the resulting genomes. Moreover palindromic sequences are notoriously difficult to propagate without deletion. Few *published* data document Rec-dependent instabilities in lambda recombinants, but such instabilities may not be uncommon, while others may be Rec-independent (and pro-

Table 2. Bacterial Hosts

Phenotype[a]	Relevant Genotype[b]	Strain[c]	Use	Reference
	sup^o	R594	non-permissive for λ *am* phage	1
	$supE$	C600	for assay and propagation of λ phage	1, 2
rk⁻	$supE$ $supF$ $hsdR$	ED8654(LE392,EQ82,BHB2600,BNN45)[d]	as above, and good transfection host	1, 2
rk⁻	$supF$ $hsdR$	NM538	for assay and propagation of λ phage	1, 6
rk⁻ mk⁻	$supE$ $hsdS$	WA803	as above, and good transfection host	1
rk⁻ Rec⁻	$supE$ $supF$ $hsdR$ $recA$	NM531	as ED8654 but Rec⁻	1, 6
rk⁻ mk⁻ Rec⁻	$supE$ $hsdS$ $recBC$	546	RecBC⁻ host for Spi⁻ phage	P.Kourilsky, personal communication
rk⁻ Rec⁻	$supE$ $supF$ $hsdR$ $recA/pgam$	GC508	RecA⁻ host for Spi⁻ phage	see 1
rk⁻ Rec⁻	$hsdR$ $recBC$ $sbcB15$ $recA$	CES201[e]	RecA⁻ host for Spi⁻ phage	C.Shurvington and F.W.Stahl (personal communication)
rk⁻ RecF⁺	$hsdR$ $recBC$ $sbcB15$	CES200[e]	RecF⁺ host for Spi⁻ phage	C.Shurvington and F.W.Stahl (personal communication)
rk⁻ RecE⁺	$hsdR$ $recBC$ $sbcA23$	NM519	RecE⁺ host for Spi⁻ phage	1
rk⁻ (P2)	$supE$ $hsdR$ (P2)	Q359	Selection of Spi⁻ phage	1, 7
rk⁻ (P2)	$supF$ $hsdR$ (P2cox)	NM539	Selection of Spi⁻ phage (derivative of NM538)	1, 6

[a]Relevant phenotypic differences from standard *E. coli* K12.

[b]*supE* suppresses most λ amber mutations other than *Sam7* and *Sam100*; *supF* suppresses most λ amber mutations other than *Pam3*.

[c]Several of these strains are being made available by the American Type Culture Collection (12301 Parklawn Drive, Rockville, MD 20852, USA), which distributes a free booklet describing its collection of Recombinant DNA Vectors and Hosts.

[d]All descended from ED8654.

[e]Strains include tn10.

bably unavoidable). Propagating the library in Rec⁻ cells should reduce homology-dependent recombination, but has the disadvantage that the yield of phage is reduced. This may be due both to a decreased burst-size, and to the relatively high proportion of dead cells in a Rec⁻ culture. Phage that adsorb to dead cells will be lost. Thus, whenever a Rec⁻ strain is used some decrease in yield must be accepted in return for a potential gain in the stability of the cloned sequences.

Genetic selection for the *red⁻ gam⁻* (Spi⁻) derivatives of EMBL vectors, on hosts lysogenic for P2, necessarily involves the use of Rec⁺ cells. Maintenance of the *recBC* product is essential for survival of the P2 lysogen (Section 14), and the formation of concatenated phage DNA is dependent upon the products of the *recA, recB* and *recC* genes (Section 13.1).

The *red⁻ gam⁺* recombinants derived from Charon 34 and Charon 35 can be propagated under the most stringent Rec⁻ conditions (*recA⁻*). The propagation of *red⁻ gam⁻* recombinants under Rec⁻ conditions is also possible, provided one is prepared to dispense with genetic selection (in which case the central fragment must be removed or otherwise rendered unable to compete for ligation to the vector arms). One way is to use a host that carries a plasmid containing the *gam* gene of lambda (G. Crouse, personal communication). This plasmid is constructed such that expression of its *gam* gene is activated by the incoming phage. In this way a *recA⁻ recBC⁺* host can be used, the *recBC* nuclease will be inactivated (Section 13.1), and replication will proceed normally. However, this *recA⁻* host cannot be used if the library is to be screened with a radioactive probe derived from a plasmid with homology to the plasmid carrying the *gam* gene. Alternatively a *recBC⁻* or a *recA⁻ recBC⁻* host is used, without recourse to a *gam* plasmid. Both strains are comparatively feeble however, the latter so much so that it will never produce high titer lysates and will, if possible, throw off more vigorous Rec⁺ revertants or pseudorevertants. It is important, therefore, to check the phenotype of this strain whenever a fresh colony is isolated (Section 18.2). If Rec⁺ conditions are permissible, and Spi-selection is dispensed with, a *recBC⁻ sbcA⁻* (RecE pathway) or *recBC⁻ sbcB⁻* (RecF pathway) host may be used. Both of these hosts are healthy and allow rolling circle replication. Neither discriminates between *gam⁺* and *gam⁻* phages, rather they encourage *gam⁻* infections to produce a large burst of phage. A *recBC⁻ sbcB⁻* host is also worth trying should there be concern for the preservation of palindromic sequences (22 and Wyman, Wolfe and Botstein in press). Taking these observations to their logical conclusion a *recA⁻ recBC⁻ sbcB⁻* host should provide the maximum safety, but it will undoubtedly limit the yield of phage particles.

17. LAMBDA OR COSMIDS

The major advantage of using a cosmid vector (2,23,24) to make a genomic DNA library resides in the greater length of DNA fragments that may be cloned (~45 kb). The benefit stems not from the modest reduction in the number of recombinants required for a representative library (*Table 1*), but from the reduction in the number of steps required to walk along a genome, and in the possibility of including in a single DNA fragment certain particularly long eukaryotic genes.

A second advantage of cosmids is the absence of large sections of the lambda genome which, being present in lambda recombinants, may add to the confusion of restriction mapping, although this must be balanced against the fact that 45 kb inserts will inevitably be more difficult to map than 20 kb inserts. It remains to be seen how much alternative methods, such as analysing partial digestion products (Section 11), will ease the mapping of phage DNA, particularly if a lambda vector were available in which the *cohesive ends* and one polylinker sequence were juxtaposed.

One disadvantage of cosmid systems is that they are not as efficient as lambda vectors in the recovery of recombinant DNA molecules. This is particularly relevant when supplies of donor DNA are limiting. It must be remembered that considerably higher molecular weight DNA will be required as starting material for efficient cloning in cosmids than will be necessary for cloning in lambda.

A second and major disadvantage is that screening by DNA hybridisation is less efficient for cosmids (colony-hybridisation) than for phages (plaque-hybridisation). This reflects in part the reduction in plasmid copy number associated with an increase in plasmid size, a problem which may be alleviated by the use of alternative cosmid systems (Little and Cross, in press). These alternative cosmid systems, in which copy number is unaffected by insert size, should also be less subject to deletion, a problem to which cosmid recombinants are notoriously prone. As an alternative to screening by hybridisation, a recent publication (25) documents the efficient selection of cosmid clones as the result of homology-dependent recombination with a plasmid encoding an additional drug resistance.

In summary, vector systems are continually evolving, but novel systems take time to be developed, assessed, and approved.

18. TECHNICAL DETAILS

18.1 Media and Buffers

18.1.1 *L-broth Media*

(i)	L-broth	Bacto-tryptone (Difco)	10 g
		Yeast extract (Difco)	5 g
		NaCl	10 g
		Water to	1 l

Adjust to pH 7.2 with NaOH and supplement with 10 mM $MgSO_4$ for the growth of lambda and its derivatives.

(ii)	L-broth agar	As L-broth with the addition of:	
		Bacto-agar (Difco)	15 g

(iii)	L-broth top layer	As L-broth with the addition of:	
		Bacto-agar (Difco) or gel quality agarose	6.5 g
		$MgSO_4$	10 mM

18.1.2 *BBL Trypticase Media*

(i)	BBL-broth	Trypticase (Baltimore Biological	
		Laboratories)	10 g
		NaCl	5 g
		Water to	1 l
		pH 7.2	
(ii)	BBL-broth agar	As BBL-broth with the addition of:	
		Bacto-agar (Difco)	10 g
(iii)	BBL top layer	As BBL-broth with the addition of:	
		Bacto-agar (Difco) or gel quality	
		agarose	6.5 g
		$MgSO_4$	10 mM

18.1.3 *Buffers*

(i)	Phage Buffer	Tris-HCl pH 7.5	50 mM
		NaCl	100 mM
		$MgSO_4$	10 mM
		gelatin	2%
(ii)	TE	Tris-HCl pH 8.0	10 mM
		EDTA pH 8.0	1 mM
(iii)	SSC	Sodium chloride	0.15 M
		Sodium citrate	0.015 M
(iv)	10xLigase		
	Buffer	Tris-HCl pH 7.5	0.675 M
		$MgCl_2$	0.1 M
		dithiothreitol	0.15 M
		spermidine	10 mM
		ATP	10 mM

18.2 **General Hints**

A number of general points, relevant to many of the procedures are given below:

(i) All buffers, other solutions, enzymes, packaging extracts, DNA preparations, gel loading buffers etc., to be used during the procedures should, where possible, be freshly prepared or obtained, and should be set aside for the purpose of library construction alone. Their integrity must be maintained at all times. Much time has been wasted in several laboratories by cloning plasmid DNA sequences that have been inadvertently introduced into batches of enzymes or other reagents. It should be noted that impurities or oxidation products present in some batches of phenol confer a facility to nick DNA. Advice on the preparation and use of phenol can be found in references 1 and 16.

(ii) Check that the vector has the expected physical and biological properties. The physical structure of the vector should be verified by restriction enzyme analysis, the biological properties by plating on the appropriate indicator strains. For example, the growth of EMBL vectors should be seriously retarded on an *E. coli* (P2) lysogen. Likewise the genetic properties of bacterial strains (rec^+/rec^-, P2/P2$^-$, r_k^+/r_k^-, m_k^+/m_k^-) should be verified (see Section 16). Recombination status can be checked by exposure to u.v. light (2). *Rec$^-$* cells are impaired in their ability to repair u.v.-induced damage to their DNA and, therefore, are more readily killed.

(iii)   Use only clean and sterile disposable plastic-ware for the precipitation of small quantities of DNA, and for the storage and transfer of enzymes, DNA solutions etc. Glass receptacles can irreversibly bind small amounts of DNA.

(iv)   Ethanol precipitation of small amounts of DNA is not always quantitative (but see reference 26). It is easy to dislodge or overlook a very small pellet of DNA if it has been over-dried. Accidents can also occur in the vacuum desiccator if the vacuum is released precipitously. The top of the tube should be covered with perforated parafilm during desiccation. N.B. High molecular weight DNAs may take some time to dissolve.

(v)   Beware of errors when pipetting very small volumes. It is extremely unlikely that different micro-pipetting devices will be found to dispense equal volumes in the $1-5$ μl range. For this reason, always use the same pipette for diluting small quantities of enzyme etc. This is particularly important when determining the activity of, and when carrying out the bulk digestion with, *Sau*3A or *Mbo*I (see Section 3.2). For measuring 1 μl volumes, a glass microcapillary is probably the most reproducible method. Be very careful when pipetting viscous DNA solutions as they take some time to enter the tip of a micropipette. For such solutions it may be advisable to cut off the end of the disposable tip.

(vi)   Sterile DNA solutions containing no active enzymes can be stored for long periods at 4°C without much harm. Freezing at -70°C may be preferable to -20°C (15). High molecular weight genomic DNA should never be frozen.

(vii)   Never put all of your eggs into one basket! Cloning in lambda involves a series of consecutive reactions. Errors can occur, sometimes for trivial reasons, at any stage. Assuming that the error can be recognised as such, which will be the case if the appropriate controls have been included, it will be a simple matter to rescue the operation as long as a fall-back is available. For this reason it is preferable to have excess starting material, and to retain a proportion of the reaction mixture at intermediate stages. The need for adequate controls cannot be over-emphasised.

(viii) A comprehensive list of recipes for growth media and buffers may be found in the Experimental Methods section of reference 1. All media should be autoclaved before use.

18.3 Checking Enzymes

Check all enzymes for the presence of the appropriate activity, and the absence of contaminating nuclease activities that might deleteriously affect subsequent ligation of DNA fragments. For most restriction enzymes that recognise hexanucleotide sequences, this is best achieved by cleaving to completion a small plasmid DNA molecule that has only one site for the enzyme (pBR322 has unique *Eco*RI, *Bam*HI, *Sal*I and *Hind*III sites). Then inactivate the enzyme, ethanol precipitate the DNA, wash the precipitate with 70% ethanol, and resuspend the DNA in ligase buffer (Section 18.1) at a concentration of ~ 100 μg/ml. Incubate half the sample with ligase and use samples of uncleaved, cleaved, and cleaved + ligated vector to transform *E. coli* to ampicillin resistance (Chapter 6). An enzyme should only be used for cloning if high efficiency transformation is restored by treatment with ligase. For most enzymes that recognise a tetranucleotide sequence, a plasmid with only one site will be more difficult to find, but φX174 DNA has only one site for *Sau*3A and *Mbo*I.

Ligation can also be monitored by running cleaved and cleaved + ligated DNA samples in adjacent tracks on an agarose gel. Successful ligation will be obvious due to the change in the mobility of the ligated (circular or multimeric linear forms) of the DNA.

Batches of phosphatase can vary considerably in quality. Phosphatase activity can be monitored by including a phosphatase step between cleavage and ligation of a small plasmid. To check that the phosphatase does not inhibit subsequent ligation, ligate a non-phosphatased donor DNA preparation (e.g., your genomic DNA cleaved by *Eco*RI) to a sample of the cleaved and phosphatased plasmid (e.g., pBR322 cleaved by *Eco*RI). Addition of the donor DNA should greatly stimulate the recovery of *amp*^R transformants derived from the phosphatase-treated plasmid.

Ensure that each batch of enzyme tested will be sufficient for the construction of the library.

18.4 Isolation of High Molecular Weight Chromosomal DNA from Eukaryotic Cells

This method is adapted from one described in reference 27.

(i)   The highest molecular weight DNA will be obtained from fresh living cells or tissue. If it is not possible to process the samples immediately, freeze them quickly by immersion in liquid nitrogen (N_2), and store them frozen at $-70°C$. Cut large pieces of tissue into small pieces with a pair of surgical scissors before immersion. This will assist both their rapid freezing and their eventual pulverisation (step ii).

(ii)   Precool a mortar and pestle with liquid N_2. Pour liquid N_2 into the mortar and add the sample (1−2 g). Before the liquid N_2 has evaporated completely, begin to grind the sample to a fine powder with the cooled pestle.

(iii)   Using a small paint brush (cooled in liquid N_2) transfer the powder from the mortar into an homogeniser, on ice, containing 10 ml of ice-cold homogenisation solution (10 mM Tris-HCl, pH 7.4; 60 mM NaCl; 10 mM EDTA; 0.15 mM spermine; 0.15 mM spermidine; 0.5% (v/v) Triton X-100). We use a Wheaton 15 ml homogeniser. Homogenise and decant the solution through fine gauze into a sterile centrifuge tube on ice. If necessary, repeat the procedure with a further 1−2 g of powdered tissue. Finally, wash the homogeniser with 5 ml of ice-cold homogenisation solution and filter into the centrifuge tube.

(iv)   As soon as possible, spin the homogenate at 7000 r.p.m. for 7 min at 4°C in a cooled rotor.

(v)   Decant the supernatant and resuspend the nuclear pellet in 10 ml of ice-cold homogenisation solution. This is easiest done if the pellet is broken up slightly in the centrifuge tube (using the homogeniser pestle) and the entire contents of the tube are returned to the homogeniser and re-homogenised.

(vi)   Repeat steps iv and v.

(vii) After the washed nuclei have been completely resuspended by homogenisation in an appropriate volume of ice-cold homogenisation solution, transfer the homogenate to a capped tube (e.g., 50 ml Falcon tube), make it 2% in Sarkosyl (from a 10% w/v stock solution), and mix by gentle inversion of the tube. Lysis occurs rapidly to produce a viscous solution. Incubate the tube at 50°C for 1 h to ensure that lysis is complete.

(viii) Weigh the lysate. Add 1.25 g of CsCl per g of lysate solution (55.55% (w/w)). Mix by gentle inversion until the CsCl has dissolved.

(ix)   Fill centrifuge tubes with this solution and centrifuge to equilibrium at room temperature (e.g., 48 h at 39 000 r.p.m. in the Beckman SW41 rotor; 24 h at 55 000 r.p.m. in the Beckman Type 65 rotor; 16 h at 45 000 r.p.m. in the Beckman VTi50 rotor).

(x)   Puncture the tube just above the bottom (where there will be an RNA pellet) with a 16 to 18 guage hypodermic needle and allow the gradient to drip out. The fractions ($0.5-1$ ml) that contain DNA will be near the middle of the gradient and will be obvious by their viscosity. Alternatively, or if the DNA is too dilute for viscosity to be apparent, spot $2-5$ μl of each fraction onto a 1% agarose gel containing 1 μg/ml ethidium bromide. Allow the spots to dry/diffuse into the agarose. Those containing DNA are recognised by their fluorescence when visualised under u.v. light.

(xi)   This step is optional but will give cleaner and more concentrated DNA solutions. Pool the appropriate fractions, dilute appropriately with a 55.55% (w/w) solution of CsCl, re-centrifuge in a smaller tube (e.g., for the Beckman SW50.1 or Beckman VTi65 rotors) and collect as described above.

(xii) Pool appropriate fractions and either:

(a) Dialyse exhaustively against TE at 4°C. The samples may be sufficiently concentrated at this stage, in which case they should be stored in sterile tubes at 4°C. If not they may be concentrated by successive gentle extractions with an equal volume of iso or 2-butanol (2). When sufficiently concentrated, dialyse the aqueous phase exhaustively against TE at 4°C to remove traces of butanol. This is probably the better method.

(b) Dilute with two volumes of TE, add twice the final volume of ethanol and mix by very gentle inversion. Threads of precipitated DNA coalesce and float to the surface as a blob. Pick this up with a plastic pipette tip and transfer it to a tube on ice. Remove any excess liquid with a micropipette and resuspend the DNA in an appropriate volume of TE. Clean DNA, which has not been overdried, will resuspend quickly. Dialyse exhaustively against TE at 4°C.

(xiii) DNA prepared in this way has proved suitable for the construction of lambda libraries and for cleavage with the majority of restriction endonucleases. If a proteinase K digestion and/or a phenol extraction step is thought to be necessary, however, these may be included between steps vii and viii or after step xii. Phenol extraction must be *gentle*. Slow rolling in a half-filled tube is probably the least injurious method.

(xiv) Determine the DNA concentration.

(xv) Analyse by electrophoresis in a 0.3% agarose gel with appropriate size markers (e.g., intact lambda DNA ~50 kb; intact bacteriophage T4 DNA ~165 kb). High molecular weight DNA will migrate as a relatively tight band behind the lambda DNA.

18.5 Partial Cleavage by *Sau*3A

(i) Mix 10 μg of genomic DNA with 10 μl of 10x *Sau*3A buffer and adjust to 100 μl with sterile distilled water.

(ii) Dispense 20 μl into a microcentrifuge tube (tube 1) and 10 μl into additional tubes (tubes 2 − 9). Cool the tubes on ice.

(iii) Commercially available preparations of *Sau*3A (and *Mbo*I) are usually supplied at 5 − 10 units per μl. Dilute the enzyme immediately before use, in the solution recommended by the supplier, to a concentration of approximately 1 unit per μl. Add 1 μl to the first tube and mix (0.5 units per μg). Remove 10 μl and mix with the contents of tube 2 (0.25 units per μg). Continue the serial dilution until tube 8, taking care to mix thoroughly at each step. Tube 9 is an undigested DNA control.

(iv) Place all tubes in a 37°C water bath and incubate for 1 h.

(v) Stop the reaction by adding EDTA to 20 mM. Cool the samples on ice, and separate them by horizontal electrophoresis overnight in a 0.3% agarose gel at 2V per cm (*Figure 2*). Low voltages improve the resolution of high molecular weight DNA species (13). Pour the 0.3% gel in a cold room and overlay the gel with electrophoresis buffer before removing the slot former. Handle the gel with care as it is quite fragile. Appropriate size markers are lambda DNA cut with *Eco*RI or *Hind*III, or one of the replacement vector DNAs cut with *Bam*HI. Stain the gel with ethidium bromide (0.5 μg/ml for 30 min), wash the gel for 30 min and photograph taking care not to over-expose the film.

(vi) Cover regions of the photograph above and below the desired size range (18 − 22 kb) and determine the concentration of enzyme that produces the maximum fluorescence in that range. Half that concentration of enzyme will produce the maximum *number* of molecules in the range (2).

(vii) Scale up the reaction to digest 150 − 300 μg of DNA. Maintain all variables as in the trial reaction (e.g., enzyme and DNA concentration, time and temperature). Pre-warm the DNA solution to 37°C before addition of the enzyme as the larger volume will take some time to equilibrate. Mixing should be gentle (no vortexing) but thorough in order that the solution be homogeneous with respect to DNA and enzyme concentrations. Choose a range of digestion conditions spanning, by a factor of two on either side, those determined as being optimum. This will help to minimise misrepresentation of fragments containing an abnormal distribution of cleavage sites. Stop the reaction by addition of EDTA to 20 mM.

(viii) Analyse an aliquot(s) of the DNA by electrophoresis in a 0.3% agarose gel, along with appropriate markers and undigested DNA, to check that the correct size-distribution has been obtained. Store the rest of the sample(s) at 4°C.

(ix) Pool the sample(s), gently extract twice with a phenol:chloroform:isoamyl alcohol mixture (25:24:1), adjust the NaCl concentration to 0.3 M, and precipitate the DNA by mixing with two volumes of ethanol. Place on ice for 10 min and then collect the DNA by centrifugation. Wash the pellet twice with 70% ethanol, and resuspend it in $200-500$ μl of TE. Resuspension may be hastened by brief (10 min) heating to 65°C.

18.6 Size Fractionation in Velocity Gradients

Sucrose-gradient fractionation is described in reference 2. Sodium chloride gradient fractionation has the advantage that a sucrose removal step is unnecessary.

(i) Pour a $5-25\%$ (w/v) NaCl gradient in a 12.5 ml centrifuge tube such as that for the Beckman SW41 rotor. Use NaCl solutions made up in 3 mM EDTA pH 8.0 and pour the gradient slowly (~ 1 ml per min), using a standard gradient former.

(ii) Layer the DNA sample in a volume of 200 μl onto the top of the gradient. Use < 150 μg of DNA per tube as larger amounts will lead to poor resolution.

(iii) Centrifuge at 37 000 r.p.m. in a Beckman SW41 rotor or its equivalent for 4.5 h at room temperature.

(iv) Collect 0.25 ml fractions in sterile Eppendorf tubes.

(v) Take 25 μl aliquots from alternate tubes, dilute with $3-4$ volumes of TE and analyse by electrophoresis in a 0.3% agarose gel, with appropriate markers (*Figure 3*). Store the rest of the samples at -20°C.

(vi) To the fractions that contain DNA of the required size-range add an equal volume of distilled water and 1 ml of ethanol. Mix well and place in a dry ice/ethanol bath for 10 min, or at -20°C overnight. Collect the DNA precipitates by centrifugation in a microcentrifuge for 15 min. Remove the supernatants, wash the pellets twice with 70% ethanol, and dry them in a vacuum desiccator. Resuspend each pellet in $20-50$ μl of sterile TE.

(vii) (Optional) Pool appropriate fractions and repeat steps (i)$-$(vi).

(viii) Spot $1-2$ μl from each fraction onto a 1% agarose gel containing 1 μg/ml ethidium bromide. Also spot serial dilutions of vector (or other) DNA, in $1-2$ μl volumes, at known concentrations that span the expected sample concentrations. Allow the DNA to dry and diffuse into the agarose. Estimate the sample concentrations by visual comparison of their fluorescence with that of the vector DNA standards.

(ix)   (Optional) Phosphatase the ends (Section 3.2 and reference 6).

18.7 Checking Vector DNA

(i)   The DNA should have the expected distribution of sites for cleavage by restriction enzymes.

(ii)   The uncleaved vector DNA must be capable of high efficiency packaging in packaging extracts of known efficiency.

(iii)   Cleavage of the vector DNA to completion with the appropriate restriction enzyme (usually *Bam*HI and used in 2- to 3-fold excess) should cause a dramatic reduction in packaging efficiency ($<10^4$-fold). N.B. Batches of enzyme may vary.

(iv)   Ligation of cleaved vector to itself should restore the phage yield although never to that obtained with the uncleaved vector DNA. The cause of inefficient ligation may be simply a defective batch of ligase or ligase buffer. On the other hand, it could be that the cohesive termini produced by the restriction enzyme and/or the *cohesive ends* of the phage DNA have been damaged at some stage. For example, it is not unknown for certain batches of restriction enzymes and ligase, more commonly the former, to carry unadvertised nuclease activities.

(v)   Certain vectors (e.g., the EMBL series) allow a genetic selection against non-recombinant (i.e., parental sequence) phage during propagation of the library (Sections 14 and 15), in which case removal of the central fragment from the cleaved vector DNA population may be unnecessary. When using these vectors it is necessary only to inactivate the restriction endonuclease before the ligation reaction. It is important, however, to assay the 'background' when the re-ligated vector is propagated on a host non-permissive for the parental phage. The level of background is always greater than zero and may vary with different batches of vector, restriction enzymes, or packaging extracts. The background must be taken into account when plating out the library for screening.

18.8 Cleavage of Vector DNA

(i)   Determine empirically the conditions for complete digestion of the vector DNA with *Bam*HI. Analyse the cleaved DNA by agarose gel (0.5%) electrophoresis.

(ii)   Digest $20-50$ µg of vector DNA with a 2- to 3-fold excess of *Bam*HI and check that the reaction has worked by electrophoresis of a sample (0.5 µg). Remember to disassociate the lambda arms by heating to 68°C for 10 min before loading. Determine the percentage of uncut phage DNA molecules by *in vitro* packaging of a sample, and re-digest if necessary. Meanwhile store the vector DNA preparation at −20°C.

(iii)   Either:
        (a) (This is appropriate for EMBL vectors in combination with a genetic

selection against parental sequence phages.) Inactivate the enzyme (step iv) and use directly for cloning.

(b) Enzymatically inactivate the central fragment of EMBL3 (6). To do this, adjust the reaction buffer to EcoRI conditions and digest with a 3-fold excess of EcoRI. Inactivate the enzymes (step iv), but precipitate with isopropanol rather than ethanol to remove the linker fragments (step v), and use for cloning.

(c) Physically remove the central fragment. Before doing this, inactivate the enzyme (step iv), adjust the vector DNA preparation to 10 mM $MgCl_2$ and incubate at 42°C for 1 h to promote the annealing of the cohesive ends (2). Check that this has occurred by electrophoresis of a sample in an agarose gel. Use as size-markers intact lambda DNA and an aliquot of cleaved vector DNA that has been heated to 68°C for 10 min to disassociate the cohesive ends. Separate the arms from the central fragment by sucrose-, NaCl-, or KI-gradient fractionation (2). Check the extent of purification by electrophoresis in a 0.5% agarose gel.

(iv) Inactivate restriction enzymes by extracting gently once with an equal volume of phenol:chloroform:iso-amyl alcohol (25:24:1), and once with chloroform:iso-amyl alcohol (24:1). Adjust the NaCl concentration to 0.3 M, and precipitate the DNA by mixing with 2 volumes of ethanol. Collect the precipitate by centrifugation in a microcentrifuge, wash it twice with 70% ethanol, dry it briefly in a vacuum desiccator, and resuspend it in TE at an appropriate concentration, say $250-500$ μg/ml.

(v) The following DNA precipitation protocol can be used to selectively remove very short-DNA fragments (e.g. linkers). Add 0.1 volume of 3 M sodium acetate, pH 6.0. Add 0.6 volumes of isopropanol, mix, and after 15 min on ice spin in a microcentrifuge for 5 min at room temperature (6). Remove the supernatant and wash the pellet with 0.5 volumes of 40% isopropanol, 0.3 M sodium acetate pH 6.0. Dry the pellet briefly in a vacuum desiccator and resuspend it in TE at an appropriate concentration.

(vi) Determine the exact DNA concentration.

18.9 Determination of the Optimal Ligation Conditions

(i) Ligate small aliquots of donor DNA, from gradient fractions that embrace the optimum insert length (15−25 kb say), to a constant amount of vector DNA. Package, and titer (see Section 13.2) on the appropriate bacterial strains. If the partial digestion conditions were chosen correctly, maximum efficiency will be coincident with fractions of the most desirable size (~20 kb for the vectors described in Section 15). The efficiency should tail off significantly with fractions larger than the maximum acceptable to the vector. Care should be taken to ensure that fractions just at the limit of acceptability are not chosen for library construction (see Section 3.2). Pool the appropriate fractions.

(ii) Ligate varying amounts of the pooled donor DNA, or of phosphatased donor DNA, to a constant amount of vector DNA (at $100-150$ μg/ml), package and titer. For a vector DNA preparation from which the central fragment has been physically removed, or enzymatically prevented from re-ligating, maximum efficiency should be obtained with roughly equal *molarities* of annealed vector arms and donor DNA fragments. So, if the vector arms (~ 30 kb) are at 100 μg/ml, try including the donor DNA fragments (20 kb average length) at expected concentrations of 15,30,60 and 120 μg/ml (67 μg/ml would give equi-molarity).

If the central fragment has not been removed (i.e., if one is relying on a genetic selection system to eliminate parental sequence phages), the arms should be present in molar excess since the efficiency of recovery of *donor* DNA should increase asymptotically as the ratio of arms to inserts increases (6). A compromise must be made, however, between maximum efficiency and keeping the size of the ligation reaction within reasonable limits.

The optimum conditions, so obtained, should be adequate for most purposes.

(iii)   At all stages of testing care should be taken to maintain constant all other variables such as ligase concentration, buffers, packaging extracts, and host strain, and to pre-anneal the *cohesive ends* (by incubation of the cleaved vector DNA at 42°C for one hour in the presence of 10 mM MgCl$_2$). Always follow exactly the instructions for handling the packaging extracts. For the purposes of titering the phage, choose a culture medium that will enable large (easily visible) plaques to be produced (Section 13.2). If the host used to propagate the library will be other than the one used to determine the optimum ligation conditions then check the efficiency with which recombinant plaques are obtained on it.

(iv)   Include as controls one packaging reaction without adding any DNA, and one with the cleaved vector DNA ligated to itself at the same concentration as used in the trial reaction. 'Background' due to uncut vector DNA molecules, contamination of the vector arms with central fragments, to 'failure' of the Spi-selection system (Section 14), or to phage DNA present in the packaging extracts, must be taken into account when scaling up. Several of these contributions to background could be assessed most directly by screening a portion of the library with a radioactive probe homologous with part of the central fragment of the vector.

(v)   Save your accumulated trial reactions towards your eventual library.

Each new batch of packaging extracts should be checked for the efficiency with which it can package a 'standard' lambda DNA preparation, such as the vector being used for cloning. It is not necessary, however, to use the highest efficiency packaging extracts for the preliminary trials described above.

18.10 A Specific Protocol for Hybridisation Screening

(i)   For either low or high density screening, plate the phage to ensure that the plaques are small and of uniform size (Section 13.2). Avoid very fresh (wet) plates, use agarose instead of agar in the top layer and invert the plates during incubation. When the plaques have appeared place the plates in a refrigerator (4°C) for at least one hour. This will discourage the top layer from being removed when filters are lifted. Filter replicas can be taken even after several weeks of storage.

(ii)   Use a pencil to number the nitrocellulose filters. Then lay them on the plates and allow them to absorb moisture from the top layer. Nitrocellulose may carry an electrostatic charge that causes it to be attracted to the agar surface. Hold the filter firmly (with gloved hands) to avoid it suddenly flapping down onto the wrong region of the plate. Gently fold the filter when laying it down so that the centre of the plate is touched first, and then gradually lower the rest of the filter taking care to expel air that might otherwise form bubbles or wrinkles under the filter. With very large filters it may help to smooth the filters gently with a gloved hand. Use sterile (autoclaved) filters if there is likely to be a long delay (>2 weeks) between lifting the filters and returning to pick a potential positive signal.

(iii)   Mark the filter and the plate asymmetrically at the outer rim by making three holes through both with an hypodermic syringe needle. The holes will remain visible in the agar enabling duplicate filters to be marked in the same way before lifting (hold the plate up to a light to see the holes through the subsequent filters).

(iv)   After an appropriate period of time (~30 sec for the first filter, 1 min for the duplicate) carefully peel off the filter with blunt-ended forceps, starting from one edge. Longer times will not be harmful. If the top agar seems to be lifting off, lower the filter and start again from another point.

(v)   Lay the filters plaque-side up for 5 min on a sheet of Whatman 3MM (or similar) paper that has been soaked in denaturing solution (0.5 M NaOH; 1.5 M NaCl). Transfer the filters for 5 min to a second sheet of 3MM paper, this time soaked in neutralising solution (0.5 M Tris-HCl, pH 8.0; 1.5 M NaCl). Then immerse the filters for 5 min in neutralising solution. Rinse the filters in 2 x SSC and place on 3MM paper to dry. Bake the filters, wrapped between circles or sheets of 3MM paper, at 80°C for two hours, preferably *in vacuo*.

When processing many filters simultaneously it may be difficult to keep to the times given above. Only the denaturation step is critical. Filters may become inordinately brittle if soaked for too long in NaOH solutions. Overbaking has a similar effect.

(vi)   Hybridise with the appropriate probe and wash under the appropriate conditions (see Reference 2). Remove excess moisture from the filters by blotting on a sheet of 3MM paper. Do not overdry the filters if there is a possibility that you may need to rewash them.

(vii)   Sandwich the damp filters between two pieces of Saran Wrap. This will keep the filters in place during autoradiography without the need to stick them down. It will then be a simple matter to remove the filters, for more extensive washing, should background be unacceptably high. Record an autoradiographic image on X-ray film (2,28). Use small adhesive labels marked with radioactive ink, and placed under the Saran Wrap, to orient the film with respect to the filters.

(viii) Mark the position of the asymmetric holes on the X-ray film. Compare duplicate filters and mark the position of plaque-shaped spots that appear in the same position on both. Go back to the plate and pick either a corresponding plaque, or a plug of agar, from the correct general region (use the wide end of a sterile Pasteur pipette). Expel the plaque or plug into 1 ml of phage buffer, add 1 drop of chloroform, and leave at room temperature for >1 h, or at $4°C$ overnight, for the phage to diffuse. Matching filters to plates will be relatively simple if 90 mm Petri dishes have been used. For larger plates there may be some degree of stretching or distortion of the filter during blotting or processing. Take an agar plug large enough to accommodate any such variation. More orientation holes may be advisable for a large filter.

(ix) More than five filters can be lifted, one after the other, from one plate. If the filters become difficult to wet, return the plates to $4°C$ for a few days for them to re-hydrate. Non-specific hybridisation problems are sometimes more severe with the first filter to be lifted.

(x) If plates are to be stored for a long time, or if an array is to be replicated using a pronged replicating block (Section 12), it may be desirable to inhibit the growth within plaques of lambda-resistant *E. coli* and other contaminating bacteria. This can be achieved, to some extent, by exposure of the agar surface to chloroform vapour. Lay a sheet of aluminium foil on the bench. Cut a circle, or appropriately-shaped piece of Whatman 3MM paper, slightly smaller than the culture dish, place it on the foil and saturate it with chloroform. Invert the plate over the paper and leave for $10-15$ min. Take care that plastic dishes do not come into contact with the chloroform, which will dissolve them, and may cause the lid to adhere when it is replaced. On occasion, droplets of condensed chloroform may collect on the agar surface, and will run when the plates are turned right-side-up. This does not seem to contribute significantly to spreading of the phage. Addition of an anti-fungal agent, such as Fungizone, to the plates may also be helpful.

(xi) When toothpicking phage from a plaque onto a new plate, or when replicating a library using a replicating block, pour a layer of top agarose containing $MgSO_4$ (10 mM) and 0.1 ml of cells per 90 mm diameter culture dish. Leave it to set for a few minutes at room temperature or at $4°C$. Lightly touch the surface with the toothpick or replicating block, and sufficient phage will have been transferred. Invert the plates and incubate them overnight at $37°C$.

(xii) Before a probe DNA can be re-used for a second round of screening, it must be denatured again. If hybridising in the presence of 50% formamide, heat the probe to $80°C$ for 10 min. Otherwise boil the probe for 5 min. Cool on ice before using.

19. ACKNOWLEDGEMENTS

We thank our colleagues in Edinburgh, Glasgow, London, and Heidelberg for advice on various aspects of the experimental methodology. Thanks are also due to John Lunec, Andrew MacCabe, Ken Murray, Shaun Scahill, and Barbara Skene who read, and commented upon, various sections of the manuscript during its many revisions. It is also a pleasure to acknowledge the patient and cheerful co-operation of Linda Kerr, Mai Wyld and Betty McCready who prepared the manuscript.

20. REFERENCES

1. Hendrix,R.W., Roberts,J.W., Stahl,F. and Weisberg,R.A. (eds.) (1983) *Lambda II*, published by Cold Spring Harbor Laboratory Press.
2. Maniatis,T., Fritsch,E.F. and Sambrook,J. (1982) *Molecular Cloning. A Laboratory Manual*, published by Cold Spring Harbor Laboratory Press.
3. Sanger,F., Coulson,A.R., Hong,G-F., Hill,D.F. and Petersen,G.B. (1982) *J. Mol. Biol.*, **162**, 301.
4. Williams,B.G. and Blattner,F.R. (1980) in *Genetic Engineering, Vol. 2*, Setlow,J.K. and Hollaender,A. (eds.), Plenum Press, New York, p.201.
5. Brammar,W.J. (1982) in *Genetic Engineering, Vol. 3*, Williamson,R. (ed.), Academic Press, London, p.53.
6. Frischhauf,A-M., Lehrach,H., Poustka,A. and Murray,N. (1983) *J. Mol. Biol.*, **170**, 827.
7. Karn,J., Brenner,S. and Barnett,L. (1983) in *Methods in Enzymology, Vol. 101*, Wu,R., Grossman,L. and Moldave,K. (eds.), Academic Press, New York and London, p.3.
8. Loenen,W.A.M. and Blattner,F.R. (1983) *Gene*, **26**, 171.
9. Maniatis,T., Hardison,R.C., Lacy,E., Lauer,J., O'Connell,C., Quon,D., Sim,D.K. and Efstratiadis,A. (1978) *Cell*, **15**, 687.
10. Fuchs,R. and Blakesley,R. (1983) in *Methods in Enzymology, Vol. 100*, Wu,R., Grossman,L. and Moldave,K. (eds.), Academic Press, New York and London, p.3.
11. Enquist,L. and Sternberg,N. (1979) in *Methods in Enzymology, Vol. 68*, Wu,R. (ed.), Academic Press, New York and London, p.281.
12. Clarke,L. and Carbon,J. (1976) *Cell*, **9**, 91.
13. Sealey,P.G. and Southern,E. (1982) in *Gel Electrophoresis of Nucleic Acids — A Practical Approach*, Rickwood,D. and Hames,B.D. (eds.), IRL Press, Oxford, p.39.
14. Benton,W.D. and Davis,R.W. (1977) *Science*, **196**, 180.
15. Davis,R.W., Botstein,D. and Roth,J.R. (1980) *Advanced Bacterial Genetics*, published by Cold Spring Harbor Laboratory Press.
16. Schleif,R.F. and Wensink,P.C. (1981) *Practical Methods in Molecular Biology*, published by Springer-Verlag Inc., New York.
17. Rodriguez,R.L. and Tait,R.C. (1983) *Recombinant DNA Techniques, an Introduction*, published by Addison-Wesley, Reading, Massachusetts.
18. Rackwitz,H.-R., Zehetner,G., Frischauf,A.-M. and Lehrach,H. (1984) *Gene*, **30**, 195.
19. Hadfield,C. (1983) in *Focus Vol. 5. No. 4*, Bethesda Research Laboratories Inc., Gaithersburg, Maryland 20877.
20. Bender,W., Spierer,P. and Hogness,D.S. (1983) *J. Mol. Biol.*, **168**, 17.
21. Horii,Z. and Clark,A.J. (1973) *J. Mol. Biol.*, **80**, 327.
22. Leach,D.R.F. and Stahl,F.W. (1983) *Nature*, **305**, 448.
23. Hohn,B. (1979) in *Methods in Enzymology, Vol. 68*, Wu,R. (ed.), Academic Press, New York and London, p.299.
24. Collins,J. (1979) in *Methods in Enzymology, Vol. 68*, Wu,R. (ed.), Academic Press, New York and London, p.309.
25. Poustka,A., Rackwitz,H.-R., Frischauf,A.-M. and Lehrach,H. (1984) *Proc. Natl. Acad. Sci. USA*, **82**, 4129.
26. Shapiro,D.J. (1981) *Anal. Biochem.*, **110**, 229.
27. Bingham,P.M., Levis,R. and Rubin,G.M. (1981) *Cell*, **25**, 693.
28. Laskey,R.A. (1980) in *Methods in Enzymology, Vol. 65*, Grossman,L. and Moldave,K. (eds.), Academic press, New York and London, p.363.
29. Karn,J., Brenner,S., Barnett,L. and Cesareni,G. (1980) *Proc. Natl. Acad. Sci. USA*, **77**, 5172.

CHAPTER 2

Constructing and Screening cDNA Libraries in λgt10 and λgt11

THANH V.HUYNH, RICHARD A.YOUNG and RONALD W.DAVIS

1. INTRODUCTION – λ VECTORS FOR cDNA CLONING

The study of eukaryotic gene structure and expression relies on the availability of cloned genes as probes. In many cases, the successful strategy toward isolating a particular eukaryotic gene has been first to isolate a DNA copy of the messenger RNA encoded by that gene (a cDNA clone). The general strategy runs as follows. A cDNA library representing the mRNA population is constructed using polyadenylated RNA extracted from the appropriate tissue or cell type. The cDNA clone of interest is then identified within the population of cDNA clones by screening the library with synthetic oligonucleotide probes, cDNA probes representing differentially expressed mRNAs, or an antibody probe. The frequency at which cDNA clones of a particular mRNA species appear in a cDNA library is generally proportional to the abundance of that species in the mRNA population. To isolate cDNA clones of rare mRNAs, it is necessary to be able to construct very large cDNA libraries representative of complex poly(A)$^+$ RNA populations. This chapter presents a simple, detailed procedure for preparing cDNA libraries containing of the order of 10^5 to 10^7 recombinants. Double-stranded cDNAs prepared by this procedure are ligated into one of two λ vectors. The use of a λ vector instead of a plasmid vector makes it possible to take advantage of the high efficiency and reproducibility of *in vitro* packaging of λ DNA as a method for introducing DNA sequences into *E. coli*. The high efficiency of cloning cDNAs into λ vectors is useful when cDNA clones of rare mRNAs are sought or when mRNA for preparing cDNAs is limited in quantity.

This chapter describes the use of two λ vectors suitable for cloning cDNAs: λgt10 and λgt11. Phage libraries cloned in λgt10 are useful for screening with nucleic acid probes. Depending on the purpose of the investigation, the nucleic acid probe may consist of any of the following: a previously cloned DNA fragment, whole genomic DNA, synthetic oligonucleotides specifying a particular amino acid sequence, RNA, or cDNA. The second vector, λgt11, is an expression vector capable of producing a polypeptide specified by the DNA insert fragment (1). Therefore, phage libraries cloned in λgt11 can be screened with antibody probes to isolate DNA sequences encoding the antigens against which the antibodies are directed (2).

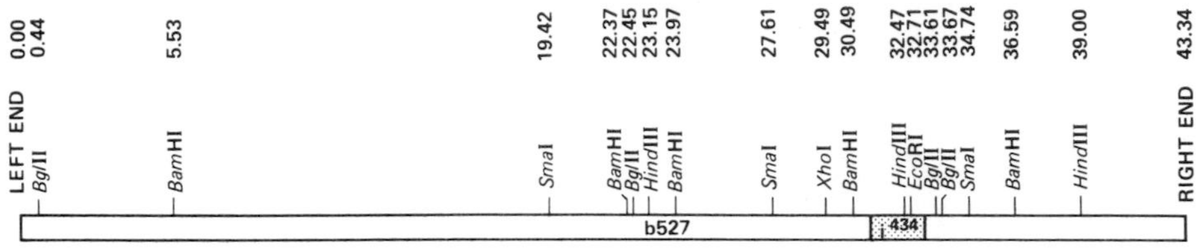

Figure 1. Map of λgt10. Restriction endonuclease cleavage sites are designated in kilobase pairs from the left end. The *b*527 deletion removes DNA sequences between 49.1% and 57.4% on the wild-type λ map. The *imm*⁴³⁴ substitution replaces DNA sequences between 72.9% and 79.3%.

1.1 λgt10

λgt10 (*imm*⁴³⁴*b*527) contains a single *Eco*RI cleavage site within the phage repressor gene. (For a map of λgt10, see *Figure 1*.) λgt10 can accept DNA insert fragments of up to 7.6 kilobase pairs, assuming a packaging capacity of 105% wild-type length. As described by Murray *et al.* (3), the insertion of a DNA fragment into the repressor gene (*c*I) generates a *c*I⁻ phage, which forms a plaque with a clear center. A *c*I⁺ phage such as λgt10 forms a turbid plaque. Recombinant *c*I⁻ phage containing insertions at the *Eco*RI site can be distinguished easily from the *c*I⁺ parent phage on the basis of their clear plaque morphology. Murray *et al.* developed several λ vectors which take advantage of this visual screen for recombinant phage and refer to them as immunity insertion vectors (3). λgt10 can be considered a member of this family of immunity insertion vectors.

Many of the previously published immunity vectors, for example NM607, NM641, NM1149, and NM1150 (3,4), are capable of being used to clone rather short DNA fragments, e.g., cDNAs. Scherer *et al.* (5) describe the successful use of *red*⁻ λ vector NM641 for the construction of cDNA libraries. λgt10 was constructed to provide a vigorously-growing λ vector optimised to accept DNA fragments a few kilobase pairs shorter than the immunity vectors available previously. This is an important consideration because λ *in vitro* packaging extracts prepared by a widely-used method package λ DNA molecules of wild-type length preferentially to shorter molecules (see Protocol II in reference 6). If the *in vitro* packaging extracts are size-selective, which is often the case, using λgt10 as a vector instead of one of its shorter relatives increases the recovery of short DNA fragments by approximately ten-fold.

Because λgt10 does not require an insert fragment to yield a packageable DNA molecule, cDNA libraries cloned in λgt10 may initially consist of 70% to 99% parent phage lacking inserts. Screening a library consisting predominantly of parent phage particles can be cumbersome. Fortunately, this problem can be eliminated by selecting against the non-recombinant parent phage during amplification of the library. The selection distinguishes between *c*I⁺*imm*⁴³⁴ and *c*I⁻*imm*⁴³⁴ phage. When a *E. coli* strain carrying the *high frequency lysogeny* mutation *hflA*150 (7) is infected by λgt10 (or any other *c*I⁺*imm*⁴³⁴ phage), the *c*I⁺ phage is repressed so efficiently that plaque formation is suppressed. However, *c*I⁻ phage form plaques with normal efficiency on the *hflA*150 strain. Plating cDNA libraries on the *hflA*150 host efficiently eliminates the parent phage from cDNA libraries cloned in λgt10. Similar results have been obtained independently by Scherer *et al.*, who plated cDNA libraries cloned in the *red⁻* λ vector NM641 on an *E. coli* host carrying the mutation *lyc*7, which is probably an allele at the *hfl* locus (5).

1.2 λgt11

The structure of the expression vector λgt11 (*lac*5 *c*I857 *nin*5 *S*100) is shown in *Figure 2*. The site used for insertion of foreign DNA is a unique *Eco*RI cleavage site located within the *lacZ* gene, 53 base pairs upstream from the β-galactosidase translation termination codon. λgt11 can accommodate up to 7.2 kb of insert DNA, assuming a maximum packageable phage DNA of 105% wild-type length. The phage vector produces a temperature-sensitive repressor (*c*I857) which is inactive at 42°C, and contains an amber mutation (*S*100) which renders it lysis-defective in hosts which lack the amber suppressor *supF*.

Because the site of insertion for foreign DNA in λgt11 is within the structural gene for β-galactosidase, foreign DNA sequences in this vector have the potential to be expressed as fusion proteins with β-galactosidase. Recombinant genomic DNA or cDNA libraries constructed in λgt11 can be screened with antibody probes for antigen produced by specific recombinant clones. In a typical experiment, 10^5 to 10^6 individual recombinants are screened in the form of phage plaques on a lawn of *lon* protease-deficient *E. coli* cells. Proteins released by the lysis of cells within the plaques are immobilised on a nitrocellulose filter placed over the lawn. Protein bound to the nitrocellulose filter is probed with antibody specific to the antigen of interest, and antibody binding is revealed in a second step by probing the filter with a radioactively-labeled second antibody or *Staphylococcus aureus* protein A. The λ recombinant responsible for the expression of the antigen of interest is isolated from the position on the plate corresponding to the signal on the filter.

In designing the λgt11 expression vector and antibody screening procedure, particular attention has been paid to problems associated with the production of foreign proteins in *E. coli*. The first problem is to achieve an adequate level of expression of the foreign DNA sequence. Having foreign DNA sequences fused

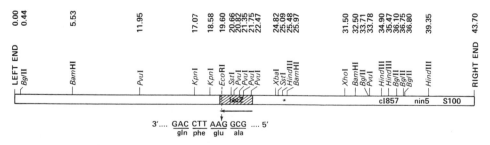

Figure 2. Map of λgt11. Restriction endonuclease cleavage sites are designated in kilobase pairs from the left end. *, λ attachment site. The transcriptional orientation of *lacZ* is indicated by the horizontal arrow. The sequence of the unique *Eco*RI site, the nucleotides that immediately surround it, and the amino acids encoded are shown below the phage map.

to β-galactosidase gene sequences ensures that the foreign DNA will be expressed efficiently in *E. coli*. The second problem is to be able to control the production of proteins encoded by the foreign DNA sequences. This is important when the foreign protein is toxic to the host cell and might kill it before sufficient amounts of antigen are produced. This problem has been minimised by using host cells producing large amounts of *lac* operon repressor (the product of the *lacI* gene) to prevent *lacZ*-directed expression of the fusion protein during the initial hours of plaque formation. When the number of infected cells surrounding the plaque is sufficiently large, *lacZ*-directed expression is induced by inactivating the repressor with isopropyl β-D-thiogalactopyranoside (IPTG). In this way, detectable amounts of the foreign antigen can be produced, even if the antigen is toxic to the cells. The third problem associated with the production of foreign proteins in *E. coli* is their stability. The position within the β-galactosidase gene chosen for fusion with foreign DNA sequences, corresponding to a region near the carboxyl terminus of the β-galactosidase protein, appears to aid in the stability of the fusion protein. This is especially noticeable in *lon* protease-deficient cells. The *lon* protease is one of several proteases responsible for the generally low stability of foreign or otherwise abnormal proteins in *E. coli* (8,9). These proteases frequently make it difficult to accumulate detectable amounts of antigen in wild-type cells. For this reason, a *lon* mutant host is used to increase the stability of these fusion proteins for the screening procedure (1).

The λgt11 system was originally designed to permit screening of expressed antigen using λgt11 recombinant lysogens (1). This approach has been abandoned in favor of screening plaques (2) for two reasons. First, many investigators have previous experience making replicas of phage plaques on nitrocellulose filters. More importantly, we have observed that some recombinant λgt11 phage repeatedly fail to integrate to form stable lysogens even in *hflA* cells (R.Young, unpublished results). This problem is not overcome by the use of λgt11-Amp3 (10). Therefore, we strongly favour screening λgt11 plaques.

The fact that λgt11 was designed as an expression vector means that it can also be used to produce a polypeptide (in the form of a β-galactosidase fusion protein) encoded by a previously-cloned DNA fragment. The hybrid protein is produced by the λgt11 recombinant clone as a lysogen of *E. coli*. Since the stability of the fusion protein is often increased in a protease-deficient host, a *lon⁻* host is also used to facilitate the isolation of preparative amounts of fusion protein from a particular λgt11 recombinant lysogen.

1.3 Factors Influencing the Decision Whether to Use λgt10 or λgt11

Because cDNA libraries cloned in λgt10 and λgt11 can both be screened with nucleic acid probes, the investigator who plans to use nucleic acid probes is presented with the choice of whether to use λgt10 or λgt11 as a vector. A major consideration relevant to this decision is that for library construction in λgt10, there is an efficient selection technique allowing only the recombinant phage to grow. A second consideration is that λgt10 is an extremely healthy bacteriophage, producing large recombinant phage plaques of uniform size and shape. These two features make λgt10 the preferred vector for the construction of cDNA libraries when nucleic acid probes are to be used to screen the library. Unlike λgt10, λgt11 was designed to express foreign DNA at high levels. During amplification of the library, some recombinants may produce low levels of polypeptides toxic to phage or host cell growth, causing a reduction in relative abundance of those recombinants in the library. In practice, some inequality in the growth of λgt11 recombinants persists even in the presence of the *lacI* repressor, which represses transcription from the β-galactosidase gene promoter. Because amplifying libraries cloned in λgt11 creates a bias in favour of healthy recombinant phage, it is recommended that λgt11 be used for the construction of libraries which will generally be screened only with antibody probes.

1.4 **Methods Included in This Chapter**

In presenting methods in this chapter, the authors' intention has been to include enough detail for the chapter to be used as an introduction to cDNA library construction and to various methods for library screening. Section 2 contains a simple, detailed procedure for synthesising and cloning cDNA in λgt10 and λgt11. In Section 3, approaches to screening λ libraries with nucleic acid probes are discussed. Much of this material is covered adequately by other methods manuals (6,11). In cases where it is not, the reader is referred to key publications which should serve as a guide. In Section 4, a method is presented for screening libraries cloned in the expression vector λgt11 with antibody probes. Finally, in Section 5, approaches to the purification of β-galactosidase fusion protein specified by a particular λgt11 recombinant clone are discussed.

2. SYNTHESISING AND CLONING DOUBLE-STRANDED cDNA IN λgt10 AND λgt11

An outline of the cDNA preparation and cloning procedure presented in this chapter is shown in *Figure 3*. Avian myelobastosis virus reverse transcriptase catalyses cDNA synthesis using poly(A)$^+$ RNA as template and oligo(dT)$_{12-18}$ as primer (12). The cDNA product is separated from the mRNA template by heat denaturation. *E. coli* DNA polymerase I catalyses the synthesis of the second strand of the cDNA, using the 3'-terminus of the first strand as a primer (12,13). Digestion with S1 nuclease removes the single-stranded hairpin region connecting the first and second strand products (12). The double-stranded cDNAs are treated with *Eco*RI methylase and S-adenosyl-methionine to methylate and protect *Eco*RI cleavage sites within the cDNAs from subsequent digestion by *Eco*RI. After a brief DNA polymerase I treatment to increase the number of flush-ended double-stranded cDNA molecules (14), *Eco*RI linkers are ligated onto the ends. Excess linkers are removed from the ends by digestion with *Eco*RI. The reaction mixture is then run over a Bio-Gel A-50m sizing column. The column serves two purposes:

(i)   to separate the cDNAs from excess linkers, which interfere with the ligation of the cDNAs to vector DNA;

(ii)   to fractionate the double-stranded cDNAs by size.

Column fractions containing double-stranded cDNAs of the desired size range

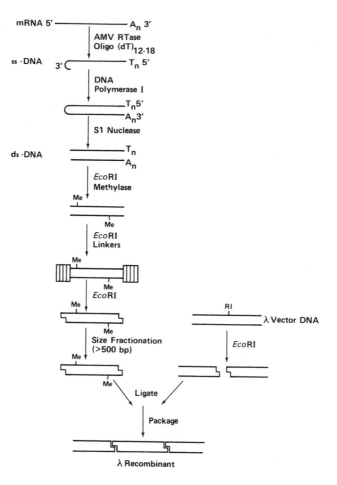

Figure 3. Scheme for double-stranded cDNA synthesis and cloning in λgt10 or λgt11. For explanation, see text.

(e.g., >500 base pairs) are pooled, ligated in a small volume with *Eco*RI-digested λgt10 or λgt11 DNA, and the ligation mixture is packed *in vitro*. A small sample of the resulting phage particles is plated out on the appropriate *E. coli* host and visually screened to determine what percentage is recombinant. If λgt10 has been used as the vector, recombinant phage are detected by their clear plaque morphology. If λgt11 has been used as the vector, recombinant phage are recognised by their property of producing colorless plaques on a *supF lacI*^{Q}-containing host in the presence of the *lac* operon inducer IPTG and the chromogenic β-galacto-sidase substrate Xgal. Once the complexity of the library has been determined, the cDNA library is plated out for amplification and for the preparation of plaque replicas on nitrocellulose filters.

Two procedures not covered in this section which are crucial to the success of the cDNA cloning procedure are the isolation of polyadenylated RNA from total RNA preparations by affinity chromatography on oligo(dT)-cellulose, and the preparation of λ *in vitro* packaging extracts. For these procedures, the reader is referred to protocols presented in reference 6. An additional reliable procedure for the preparation of *in vitro* packaging extracts appears in reference 5.

2.1 Enzymes

(i)   Avian myeloblastosis virus reverse transcriptase: Seikagaku America, Inc.

(ii)   *E. coli* DNA polymerase I: prepared from *E. coli* 594 lysate as in reference 11, or New England Biolabs.

(iii)   T4 DNA ligase: prepared from *E. coli* E1150 lysate as in reference 11, or New England Biolabs.

(iv)   Nuclease S1: Boehringer Mannheim Biochemicals.

(v)   Restriction endonuclease *Eco*RI: prepared from *E. coli* RY13 as in reference 15, or New England Biolabs.

(vi)   *Eco*RI methylase: prepared from *E. coli* RY13 as in reference 16, or New England Biolabs.

(vii)   T4 polynucleotide kinase: Pharmacia P-L Biochemicals, Inc.

Commercial sources for enzymes and references for methods for preparing them are provided to serve as a guide. Other commercial sources or methods of preparation may be superior.

2.2 Other Reagents

(i)   2'-Deoxynucleoside 5'-triphosphates (dATP, dCTP, dGTP, dTTP): Pharmacia P-L Biochemicals, Inc. Make up concentrated stocks of nucleoside triphosphates in sterile distilled water, neutralise to pH 7−8 with NaOH, and store at −70°C. Determine precise concentrations of nucleotide stocks spectrophotometrically.

(ii)   Adenosine 5'-triphosphate, disodium salt: Pharmacia P-L Biochemicals, Inc. For preparation of stock solution, see (i) above.

(iii)   Oligo(dT)$_{12-18}$: Pharmacia P-L Biochemicals, Inc. Dissolve in sterile distilled water at a concentration of 100 μg/ml and store at −70°C. Determine the precise concentration of the solution spectrophotometrically.

(iv)   S-adenosyl-L-methionine, iodide salt, grade I: Sigma Chemical Company. Make up a 10 mM stock in 10 mM sodium acetate buffer, pH 5.0, and store at −20°C. Dilute one hundred-fold to 100 μM in distilled water just before use.

(v)   *Eco*RI linkers, 5'-hydroxyl-d(GGAATTCC): Collaborative Research, Inc. Make up a 500 μg/ml stock in 10 mM Tris-HCl, pH 7.5, 1 mM Na₃EDTA and store at −20°C.

(vi)   Deoxycytidine 5'-[α-^{32}P]triphosphate, 10 mCi/ml, ~800 Ci/mmol, aqueous solution: Amersham.

(vii)   Adenosine 5'-[γ-^{32}P]triphosphate, 10 mCi/ml, >3000 Ci/mmol, aqueous solution: Amersham.

(viii)   Isopropyl β-D-thiogalactopyranoside (IPTG): Sigma Chemical Company. Prepare a 1 M solution in sterile distilled water. Store at −20°C.

(ix)   5-Bromo-4-chloro-3-indoly-β-D-galactopyranoside (Xgal): Boehringer

Mannheim Biochemicals. Make up a 40 mg/ml solution in dimethylformamide and store at −20°C.

(x) Bio-Gel A-50m, 100−200 mesh: Bio-Rad Laboratories.

Reagents (viii) and (ix) are only needed for cloning in λgt11.

2.3 Bacterial and Phage Strains

2.3.1 *Strains Required for Cloning in* λgt10

(i) BNN93 = *E. coli hsdR*$^-$ *hsdM*$^+$ *supE thr leu thi lacY*1 *tonA*21.

(ii) BNN102 = BNN93 *hflA*150[chr::Tn*10*]. The construction of BNN102 is described in reference 1.

(iii) λgt10 = λ*srI*λ1° *b*527 *srI*λ3° *imm*^{434} (*srI*434$^+$) *srI*λ4° *srI*λ5°. λgt10 was constructed by *in vitro* techniques using the left arm of NM518 (17) and the right arm of NM607 (3). Equal masses of NM518 and NM607 DNAs were mixed together, digested with *Xho*I, ligated with T4 DNA ligase, then packaged *in vitro*. One-fourth of the resulting phage were expected to have the desired genotype. Phage DNAs were screened for one possessing the expected restriction map.

Strains required for cloning in λgt10 can be obtained from Dr. Ronald W.Davis, Department of Biochemistry, Stanford University School of Medicine, Stanford, CA 94305, USA.

2.3.2 *Strains Required for Cloning in* λgt11

(i) BNN97 (American Type Culture Collection no. 37194) = BNN93(λgt11). The genotype of λgt11 is *lac*5 △(*shind*IIIλ2-3) *sri*λ3° *c*I857 *srI*λ4° *nin*5 *srI*λ5° *S*am100. The construction of λgt11 is described in reference 1.

(ii) Y1088 (ATCC no. 37195) = *E. coli* △*lacU*169 *supE supF hsdR*$^-$ *hsdM*$^+$ *metB trpR tonA*21 *proC*::Tn*5* (pMC9). pMC9 = pBR322-*lacI*^{Q}.

(iii) Y1089 (ATCC no. 37196) = *E. coli* △*lacU*169 *proA*$^+$ △*lon araD*139 *strA hflA*150[chr::Tn*10*] (pMC9).

(iv) Y1090 (ATCC no. 37197) = *E. coli* △*lacU*169 *proA*$^+$ △*lon araD*139 *strA supF*[*trpC*22::Tn*10*] (pMC9).

The construction of Y1088, Y1089, and Y1090 is described in reference 2. Strains required for cloning in λgt11 are available through the American Type Culture Collection, 12301 Parklawn Drive, Rockville, MD 20852, USA, or through Clontech Laboratories, 922 Industrial Avenue, Palo Alto, CA 94303, USA. Clontech Laboratories also distribute a number of λgt11 recombinant DNA libraries and positive controls for immunoscreenings. *Eco*RI-cleaved and dephosphorylated λgtll arms are available through Promega Biotec, 280 S. Fish Hatchery Road, Madison, WI 53711, USA.

2.4 Directions for Growth of λ Phage Vectors and Preparation of λ Vector DNA

2.4.1 *Preparation of Plating Cells*

Prepare BNN93, BNN102, and Y1088 cells for plating phage as described by Davis *et al.* (11) by growing cultures to stationary phase in LB medium (pH 7.5)/0.2% maltose. Pellet the cells and resuspend in 1/2 the original culture volume of 10 mM $MgSO_4$. Store the cells at 4°C. Use freshly prepared plating cells for plating out phage libraries.

2.4.2 *Directions for Growth of* λgt10

The following procedure is adopted to minimise the number of spontaneous clear-plaque phage in the vector preparation. Single plate stocks are grown from individual plaques and titered. The plate stock containing the lowest percentage of clear-plaque phage is chosen as a starter stock for a large-scale phage preparation. When growing up phage for the preparation of vector DNA, it is inadvisable to start with a plate stock containing more than one clear-plaque phage in 10^4, because these phage will show up as false recombinants in the library.

(i)   To plaque purify the phage, streak out the phage on a fresh LB plate (pH 7.5) as described by Davis *et al.* (11), using BNN93 as host.

(ii) Using 100 μl glass micropipettes, remove several individual turbid plaques in agar plugs and grow up individual plate stocks from them using fresh LB plates (pH 7.5).

(iii) Titer each plate stock to identify the one containing the lowest percentage of contaminating clear-plaque mutant phage.

(iv) Using the plate stock that has the lowest percentage of clear-plaque phage as the starter stock, prepare ten 150 mm plate stocks as described by Davis *et al.* (11), except use LB plates (pH 7.5) adjusted to 0.2% glucose after autoclaving instead of regular LB plates. The addition of glucose to the medium increases the titer of λgt10 plate stocks by ten-fold. Use 2.5 x 10^6 plaque-forming units (p.f.u.), $125-250$ μl BNN93 plating cells, and 5 ml LB soft agar (pH 7.5) per 150 mm Petri dish. After $5-6$ h of incubation at 37°C, the plates should have a mottled-looking lawn on them. The plates will never become completely clear. The lawn will become more turbid if left to grow and the titer of the plate stock will drop.

(v)   Cool the plates at 4°C at the point when the lawn looks the lightest. When the plates have cooled thoroughly, overlay the plates with 10 ml cold λ diluent (4°C) plus a few drops of chloroform. (λ diluent is 10 mM Tris-HCl pH 7.5, 10 mM $MgCl_2$, 0.1 mM Na_2EDTA.) Let the phage diffuse into the λ diluent at 4°C overnight.

(vi) Remove the overlay solution, taking care not to remove any agar.

(vii) Sediment the bacterial debris by centrifugation in a Beckman JA-20 rotor (or equivalent) at 10 000 r.p.m. at 4°C for 15 min.

(viii) Pellet the phage by centrifugation in a Beckman SW27 rotor at 23 000 r.p.m. at 4°C for 90 min (two buckets).

(ix) Resuspend each phage pellet in 1 ml λ diluent.

(x)   Transfer the phage suspension to 1.5 ml microcentrifuge tubes and spin in a microcentrifuge for 5 seconds to pellet the debris.

(xi) Use the supernatant from step (x) for CsCl block gradients as described by Davis *et al.* (11). Step the phage down once and up once. Run two gradients in Beckman SW50.1 tubes to purify phage from ten 150 mm plate

stocks. (For the block gradients, use analytical grade CsCl.)
(xii)   Store the CsCl-purified phage suspension at 4°C.
(xiii) Prepare λ DNA by formamide extraction as described by Davis *et al.* (11).
(xiv)   The yield of λgt10 DNA from ten 150 mm plate stocks is generally
        0.5 − 1 mg.

2.4.3 *Directions for Growth of* λgt11

λgt11 phage is prepared by inducing the λgt11 lysogen BNN97 at 43°C, a
temperature at which the temperature-sensitive phage repressor (*c*I857) is inactive.
(i)   Streak out BNN97 for single colonies at 32°C on an LB plate (pH 7.5).
(ii)   To test the lysogen for temperature sensitivity at 42°C, array cells from
        a few single colonies using sterile toothpicks onto two LB plates (pH 7.5).
        Incubate the first plate at 42°C and the second at 32°C.
(iii)   Having confirmed temperature sensitivity of growth at 42°C, inoculate 10 ml
        of LB (pH 7.5) with a single colony and incubate the culture overnight
        at 32°C.
(iv)   Use the 10 ml overnight culture to inoculate one liter of LB (pH 7.5) supple-
        mented with 10 mM $MgSO_4$ at 32°C. Grow the culture with good aeration
        at 32°C to OD_{600} = 0.6.
(v)   Increase the temperature of the culture to 43 − 44°C as rapidly as possible
        (e.g. in a pre-warmed water bath) and incubate with good aeration at
        43 − 44°C for 15 min.
(vi)   Allow the temperature of the culture to drop to 37 − 38°C and continue
        incubating with good aeration for three hours. (Do not let the temperature
        of the culture drop below 37°C.)
(vii)   Add 10 ml of chloroform to the culture and mix well for 10 min at 37°C.
        The cell suspension should become clear.
(viii) Remove bacterial debris from the culture by centrifugation at 6000 r.p.m.
        in a Beckman JA-10 rotor (or equivalent) for 10 min at 4°C.
(ix)   Harvest the phage by centrifugation at 8000 r.p.m. in a Beckman JA-10
        rotor (or equivalent) for 8 h at 4°C.
(x)   Gently resuspend the phage pellet in 4 ml of λ diluent with the aid of a
        Pasteur pipette. (λ diluent is 10 mM Tris-HCl pH 7.5, 10 mM $MgCl_2$,
        0.1 mM Na_2EDTA.)
(xi)   Add an equal volume of λ diluent saturated with CsCl. (Use analytical grade
        CsCl.)
(xii)   To band the phage, centrifuge the phage at 35 000 r.p.m. in a Beckman
        SW50.1 rotor at 20°C for 48 h.
(xiii) Remove the phage band through the side of the tube using a syringe equipped
        with a 25 gauge needle. The phage band should be near the center of the
        tube. Remove the phage in a volume of approximately 0.5 ml.
(xiv)   Add an equal volume of λ diluent.
(xv)   Further purify the phage on CsCl block gradients in a Beckman SW50.1
        rotor as described by Davis *et al.* (11). Step the phage down once and
        up once.

(xvi)   Remove the final phage band in as small a volume as possible. Store the phage suspension at 4°C.

(xvii) Prepare phage DNA by formamide extraction as described by Davis *et al.* (11).

(xviii) The yield of phage DNA from one liter of culture is generally 0.25−0.5 mg.

2.5 Procedure for Synthesising and Cloning Double-Stranded cDNA in λgt10 and λgt11

The protocols for the synthesis and cloning of double-stranded DNA in these vectors are given in *Tables 1−9* together with the parallel text of Sections 2.5.1−13 which contains additional explanatory information.

2.5.1 First Strand cDNA Synthesis

The procedure for synthesising the first strand of cDNA is given in *Table 1*. It is most important that all the reagents and plastic ware used for the first strand cDNA synthesis are free of contaminating RNase. This includes the reverse transcriptase. In order to assay the reverse transcriptase for non-specific RNase activity. Incubate a few units of the enzyme with 0.25 μg of total RNA in 50 mM Tris-HCl pH 7.5, 100 mM NaCl, 8 mM $MgCl_2$ at 37°C for one hour, then run the RNA on a 1% agarose gel in Tris-acetate buffer containing 0.5 μg/ml ethidium bromide. Examine the RNA for degradation using u.v. illumination. It may be necessary to purify the reverse transcriptase from contaminating RNase by passing it over a Sephacryl S-200 column as described by Wahl *et al.* (18).

Secondly, it is necessary that the first strand cDNA synthesis reaction is done with an excess of enzyme over template. Titrate the enzyme using a fixed

Table 1. First Strand cDNA Synthesis.

1.	Make up a 50 μl reaction mixture by adding the reagents in the order listed in this protocol into a sterile 1.5 ml microcentrifuge tube. First add a volume of water so that the volume will be 34 μl when the poly(A)$^+$ RNA and reverse transcriptase have been added.
2.	Add 2 μg[a] poly(A)$^+$ RNA in distilled water. Heat the mixture at 70°C for 3 min and cool on ice.
3.	Add in the following order: 5 μl of 10 x First Strand Buffer[b], 5 μl of 10 mM d(ACGT)TP[c], 5 μl of 100 μg/ml oligo(dT)$_{12-18}$, 1 μl of 5'-[α-^{32}P]dCTP, 10−20 units[d] of AMV reverse transcriptase to give a reaction volume of 50 μl.
4.	Incubate at 42°C for 1 h.
5.	Punch a hole in the tube cap with a needle. Denature the RNA:cDNA duplexes by heating the mixture in a boiling water bath for 1.5 min, and then cool on ice. Replace the punctured cap with a new cap.
6.	Before going on to the steps in *Table 2*, put aside 0.5 μl of the reaction product to ethanol precipitate and run on an alkaline gel to analyse the size distribution of the reaction product (see Section 2.5.3).

[a]If other RNA concentrations are used for the first strand reaction, then the RNA:oligo(dT) ratios should be held constant.
[b]10 x First Strand Buffer is 0.5 M Tris-HCl pH 8.5 (measured at room temperature), 0.4 M KCl, 0.1 M $MgCl_2$, 4 mM DTT.
[c]d(ACGT)TP is an aqueous solution containing 10 mM each dATP, dCTP, dGTP, and dTTP, according to directions given in Section 2.2.
[d]A unit of reverse transcriptase catalyses the incorporation of one nanomole of dTMP into acid insoluble product in 10 min at 35°C, using polyriboadenylate:deoxythymidylate as template-primer.

poly(A)$^+$ RNA concentration. Assay cDNA synthesis by determining the number of radioactive counts incorporated into acid precipitable material. After having established conditions in which enzyme is in excess, check to make sure that the amount of cDNA synthesised (by determining acid-precipitable counts) is linear with respect to the amount of poly(A)$^+$ RNA added to the reaction and that there is no incorporation of label when RNA is left out of the reaction.

Five to fifty per cent of the mass of the poly(A)$^+$ RNA should be reverse transcribed into DNA. This can be estimated from the number of radioactive counts incorporated into DNA during the first strand synthesis reaction as the specific activity of the label is known. If the incorporation is poor, it may be because:

(i)   the template RNA is badly degraded; or

(ii)   there is an inhibitor of the enzyme present in the RNA preparation. If a suitable gene probe is available, degradation of the poly(A)$^+$ RNA can be detected with the most sensitivity by fractionating the RNA on a gel, transferring it to a solid substrate, and hybridising the RNA with the radioactively-labeled DNA probe ('Northern' analysis). An inhibitor can be detected by mixing the RNA with another poly(A)$^+$ RNA preparation that is known to be a good template for reverse transcriptase and observing how the presence of the former RNA affects the template activity of the latter.

Finally, check the dependence of the first strand synthesis reaction on the oligo(dT) primer. Oligo(dT) should enhance the incorporation of radioactive label into DNA at least ten-fold. A high level of oligo(dT)-independent incorporation is sometimes caused by self-primed cDNA synthesis from DNA and/or residual ribosomal RNA contaminating the poly(A)$^+$ RNA preparation. Reverse transcription of these molecules results in products which migrate as discrete bands on an alkaline agarose gel. Oligo(dT)-independent incorporation due to these contaminants can sometimes be reduced by further purification of the poly(A)$^+$ RNA. A second cause of oligo(dT)-independent incorporation is associated with the reverse transcriptase preparation. We have observed that a commercial source of reverse transcriptase additionally purified on a Sephacryl S-200 column as described in reference 18 gave a much higher level of oligo(dT)-independent incorporation before the column than after. The cDNA product of this oligo(dT)-independent reaction was heterogeneous in length, unlike the reaction products resulting from contaminating DNA or ribosomal RNA. It is possible that the reverse transcriptase was contaminated by small RNA or DNA fragments which could randomly prime cDNA synthesis on the exogenously-added RNA, and that the enzyme was purified away from the small fragments on the sizing column. This is an additional reason for the further purification of reverse transcriptase on a Sephacryl S-200 column other than to remove RNase as described above.

2.5.2 *Second Strand cDNA Synthesis*

The protocol for this step is contained in *Table 2*.

Table 2. Second Strand cDNA Synthesis.

1.	To the tube containing 50 μl first strand reaction product (from Step 5 of *Table 1*), add 50 μl of 2 x Second Strand Buffer[a], 1 μl of 10 mM d(ACGT)TP, 5 μl of 1 mg/ml homogeneous DNA polymerase I or 90 units[b]. The final reaction volume is 106 microliters.
2.	Incubate the reaction mixture at 15°C for 5 h.
3.	Before going on to the steps in *Table 3*, put aside 1 μl of the reaction product to run on a gel (see Section 2.5.3).

[a]Second Strand Buffer is 100 mM HEPES adjusted to pH 6.9 with KOH, 100 mM KCl, 20 mM $MgCl_2$.
[b]A unit of DNA polymerase I catalyses the incorporation of 10 nanomoles of nucleotide in 30 min at 37°C, using poly dAT as template-primer.

Table 3. Nuclease S1 Digestion.

1.	To the tube containing 106 μl of the second strand synthesis product (from Step 2 of *Table 2*), add 400 μl of S1 Buffer[a] and about 600 units of nuclease S1[b] to give a final reaction volume of about 506 μl.
2.	Incubate the reaction mixture at 37°C for 30 min.
3.	Add 5 μl of 1 M Na_4EDTA and then put aside 5 μl of the mixture to analyse by alkaline gel electrophoresis[c].
4.	Extract the reaction mixture with an equal volume of phenol:$CHCl_3$ (1:1) saturated with TE 7.5[d]. Spin for 3 min in a microcentrifuge at room temperature to separate the phases.
5.	Remove the aqueous phase to another 1.5 ml microcentrifuge tube. Extract the aqueous phase three times with an equal volume of ether saturated with TE 7.5. Let the residual ether evaporate for a few minutes at 37°C.
6.	Add two volumes of ethanol. Incubate in a dry ice/ethanol bath for 1 h, followed by an incubation at −20°C for 24 h.
7.	Sediment the precipitate by spinning for 15 min in a microcentrifuge at 4°C. Carefully remove the supernatant with a Pasteur pipette. Let the pellet air dry.

[a]S1 Buffer is 30 mM sodium acetate pH 4.4, 250 mM NaCl, 1 mM $ZnCl_2$.
[b]The amount of nuclease S1 should be determined as described in the text of Section 2.5.3. Nuclease S1 purchased from Boehringer Mannheim in solid form is reconstituted at 1000 units/ml. A unit of nuclease S1 catalyses the formation of 1 μg of acid-soluble deoxyribonucleotides after 30 min of incubation at 37°C with single-stranded DNA.
[c]The procedure for alkaline gel electrophoresis is described in the text of Section 2.5.3.
[d]TE 7.5 is 10 mM Tris-HCl pH 7.5, 1 mM Na_3EDTA.

2.5.3 *Nuclease S1 Digestion*

This procedure is given in *Table 3*. It is necessary to titrate the nuclease S1 beforehand to determine the optimal amount to digest the hairpin structures at the ends of the double-stranded cDNAs. To assay the S1, incubate two-fold dilutions of the enzyme in the range of 1200 units/ml with radioactively-labeled second strand synthesis reaction product, then analyse the digestion products on an alkaline agarose gel. Choose the S1 concentration which reduces the size distribution of the cDNA smear to that of the first strand synthesis reaction product (see *Figure 4*).

After carrying out the reaction the size of an aliquot of the product can be determined by electrophoresis on an alkaline gel together with the aliquots saved from the steps in *Tables 2* and *3*. Add one microgram of carrier DNA and to these aliquots and ethanol precipitate the reaction products in order to remove most of the unincorporated 5'-[α-^{32}P]dCTP. Run the three samples in adjacent

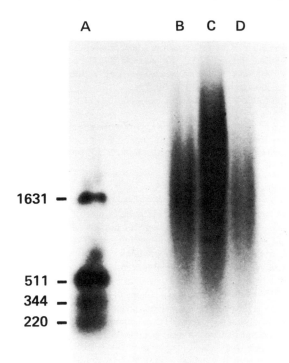

Figure. 4. Size distribution of cDNAs. Samples of ³²P-labeled cDNA were electrophoresed on a 1.2% agarose gel in alkaline buffer, which was then dried and exposed to X-ray film. **A.** Size standards in number of nucleotides. **B.** First strand cDNA product. **C.** Second strand cDNA product. **D.** cDNA after digestion with S1 nuclease.

tracks next to size standards on a 1.2% agarose gel prepared in 50 mM NaCl, 2 mM EDTA and run submerged in 30 mM NaOH, 2 mM EDTA. Dry the gel and expose it to X-ray film. An autoradiogram of a gel on which samples of first strand, second strand, and S1 digestion products have been run is shown in *Figure 4*. The size distribution of the first strand reaction product should directly reflect the size distribution of the RNA template. After the second strand synthesis, the average size of the product should have roughly doubled. After the S1 digestion, the size distribution of the product should be the same as that of the first strand reaction products.

2.5.4 *EcoRI Methylation Reaction*

The *Eco*RI methylation reaction protects the cDNA from *Eco*RI restriction when cohesive ends are generated on the synthetic linkers (see section 2.5.8). The protocol for *Eco*RI methylation is given in *Table 4*.

2.5.5 *DNA Polymerase I Fill-in Reaction*

This reaction increases the number of blunt-ended DNA molecules to which the chemically synthesised linkers can be added using T4 DNA ligase. The procedure is given in *Table 5*.

Table 4. *Eco*RI Methylation Reaction.

1.	Resuspend the pellet from step 7 of *Table 3* in 20 μl *Eco*RI Methylase Buffer[a].
2.	Add 2 μl of 100 μM S-adenosyl-L-methionine[b], 0.2 μl of 1.8 mg/ml *Eco*RI methylase or enough enzyme to methylate one microgram of DNA to give a final reaction volume of 22 μl.
3.	Incubate at 37°C for 15 min, followed by 70°C for 10 min.

[a]*Eco*RI Methylase Buffer is 50 mM Tris-HCl pH 7.5, 1 mM Na_3EDTA, 5 mM DTT.
[b]Make up the 100 μM solution of S-adenosyl-methionine according to the directions given in Section 2.2.

Table 5. DNA Polymerase I Fill-In Reaction.

1.	To the tube containing 22 μl of *Eco*RI-methylated cDNA from the steps described in *Table 4*, add 2.5 μl of 0.1 M $MgCl_2$, 2.5 μl of 0.2 mM d(ACGT)TP[a], 0.25 μl of 1 mg/ml homogeneous DNA polymerase I or 5 units[b] giving a 27 μl reaction volume.
2.	Incubate at room temperature for 10 min and then add 10 μl of 50 mM Na_3EDTA.
3.	Add one microgram of uncut λ vector DNA (λgt10 or λgt11) as carrier. Extract the mixture with an equal volume of phenol:$CHCl_3$ saturated with TE 7.5[c]. Spin for three minutes in a microcentrifuge at room temperature.
4.	Remove the aqueous phase to another 1.5 ml microcentrifuge tube and re-extract the organic phase twice with 25 μl TE 7.5. Pool the aqueous phases and extract three times with an equal volume of ether saturated with TE 7.5.
5.	After removing the final ether phase, let the residual ether evaporate for a few minutes at 37°C. Add sodium acetate to a final concentration of 0.3 M. Add two volumes of ethanol. Allow the DNA to precipitate by incubating in a dry ice/ethanol bath for 30 min.
6.	Sediment the precipitate by spinning in a microcentrifuge at 4°C for 15 min. Remove the supernatant and wash the pellet once with −20°C 70% ethanol. Centrifuge again for 10 min.
7.	Carefully remove the ethanol supernatant and let the pellet air dry.

[a]See footnote [c] to *Table 1* for a description of d(ACGT)TP.
[b]See footnote [b] to *Table 2* for a definition of the units of DNA polymerase I.
[c]See footnote [d] to *Table 3* for TE 7.5.

Table 6. 5′ End-Labeling of *Eco*RI Linkers.

1.	Combine in the following order: 34 μl of distilled water minus the volume of the kinase, 5 μl of 10 x Kinase Buffer[a], 10 μl of *Eco*RI linker[b] solution, 1 μl 5′-[γ-^{32}P]ATP and 30 units[c] T4 polynucleotide kinase to give a reaction volume of 50 μl.
2.	Incubate at 37°C for one hour.
3.	Store at −70°C. Before using the linkers, check the ability of the linkers to be ligated together as described in the text of Section 2.5.6.

[a]10 x Kinase Buffer is 0.5 M Tris-HCl pH 7.5, 0.1 M $MgCl_2$, 0.1 M DTT, 10 mM ATP.
[b]The linker solution is made up at 500 μg/ml in TE 7.5.
[c]One unit of kinase catalyses the transfer of one nanomole phosphate to the 5′-OH end of a polynucleotide from [γ-^{32}P]ATP in 30 min at 37°C.

2.5.6 5′ End-Labeling of EcoRI Linkers

The protocol for labeling the linkers is given in *Table 6*. In order to check whether the kinasing of the linkers has been successful, a ligation reaction can be carried out and the products examined by gel electrophoresis and autoradiography.

(i) Add 1 μl of the ^{32}P-kinased linkers (100 nanograms) to 4 μl 10 mM Tris-HCl pH 7.5, 10 mM $MgCl_2$, 10 mM DTT, 1 mM ATP.

Table 7. Addition of *Eco*RI Linkers to the Double-Stranded cDNA.

1.	Resuspend the pellet from Step 7 of *Table 5* in 4.5 μl of 100 μg/ml kinased *Eco*RI linkers from Step 3 of *Table 6*.
2.	Add 0.5 μl of 10 mM ATP, 0.5 μl of 1.6 mg/ml T4 DNA ligase[a] to give a reaction volume of 5.5 μl.
3.	Incubate at 12 − 14°C for one to two days.
4.	To the reaction product add 5 μl of 50 mM Tris-HCl pH 7.5, 10 mM MgSO₄, 200 mM NaCl and then heat the mixture at 70°C for 10 min to inactivate the ligase.
5.	Add 1 μl of *Eco*RI (20 units[b]/μl) to give a reaction volume of 11.5 μl and incubate at 37°C for 1 h.
6.	Stop the reaction by adding 0.2 μl of 1 M Na₄EDTA.
7.	Fractionate the double-stranded cDNA over a Bio-Gel A-50m column as described in Section 2.5.8 to remove the excess linkers.

[a]Do not add enzyme to more than 10% of the reaction volume, or else the glycerol concentration may be too high.
[b]A unit of *Eco*RI will digest 1 μg of pBR322 DNA to completion in 1 h at 37°C.

(ii)   Add 0.5 μl of 1.6 mg/ml T4 DNA ligase and incubate at 12 − 14°C for 6 − 24 h.

(iii) Run the ligation product on a 10% polyacrylamide gel next to an equal amount of unligated ^{32}P-linkers until the bromophenol blue marker is halfway down the gel. Expose the gel to X-ray film.

Most of the label should be in a ladder of linker oligomers.

2.5.7 *Addition of EcoRI Cohesive Termini onto the cDNA*

This step has two stages. The first is the ligation of the synthetic linkers to the double-stranded cDNA. Since this results in the covalent attachment of oligomers of the linkers to the ends of the cDNA, it is subsequently necessary to cleave the reaction product with *Eco*RI to remove the excess linkers. These procedures are given in *Table 7*.

2.5.8 *Removal of Excess EcoRI Linkers and Size Fractionation of Double-Stranded cDNAs*

The cDNAs are fractionated by size and excess linkers are removed in a single step by running the reaction mixture over a Bio-Gel A-50m column at room temperature. The column is made of a piece of glass tubing 32 cm long with an inner diameter of 0.2 cm. Plug the column at the bottom with a small disk of porous polyethylene or with glass wool. In a fume hood, silanise the column by washing it for a few minutes with 1% dimethyldichlorosilane in CHCl₃ followed by ethanol. Attach a 25 gauge syringe needle to the bottom of the column with Parafilm. Insert the needle into a length of PE-20 tubing sufficiently long to be used with a fraction collector. (PE-20 polyethylene tubing with an inner diameter of 0.38 mm and an outer diameter of 1.09 mm is manufactured by Clay-Adams.) Attach an empty disposable plastic syringe to the top of the column, using Tygon tubing as an adaptor, to serve as a reservoir. Wash the pre-swollen Bio-Gel A-50m resin (100 − 200 mesh) several times in Column Buffer (10 mM Tris-HCl pH 7.5, 100 mM NaCl, 1 mM Na₃EDTA). To pour the column, first inject enough buffer to fill the column through the bottom. Pour the resin into the reservoir, then let the column flow. The final bed volume is about 1 ml. Wash the column thoroughly with about 50 ml Column Buffer. This is very important to get rid of a ligation inhibitor in the resin.

Precalibrate the column by running radioactively-labeled DNA restriction fragments of known length (e.g., 5′-end labeled *Hin*fI fragments of pBR322) through the column and assaying the fractions by gel electrophoresis. To such radioactive restriction fragments in 10 μl of TE 7.5, add 1 μl of 0.25% bromophenol blue-50% glycerol. Load the sample onto the column. Adjust the flow rate so that the dye marker moves about 1 cm in 10 min. Collect 3-drop fractions in 1.5 ml microfuge tubes ($\sim$15 μl per drop). (A Gilson Micro Fractionator is a convenient fraction collector to use.) Run samples of each fraction on a 1.2% agarose gel. Dry the gel and expose it to X-ray film. Wash the column thoroughly with Column Buffer before using it to fractionate cDNAs. Once calibrated, the column can be reused many times.

To prepare the cDNA sample for the column, add 1 μl 0.25% bromophenol blue-50% glycerol to the cDNA sample. Spin the sample for 30 seconds in a microfuge to pellet any insoluble debris that might interfere with the flow of the column. Load the sample onto the column and adjust the flow rate. Collect 3-drop fractions as before. Count the radioactive label in the fractions by capping the tubes, placing them inside scintillation vials, and counting the entire sample on a ^{3}H channel. An example of an elution profile of cDNAs run over a Bio-Gel A-50m column is shown in *Figure 5*.

An alternative to precalibrating the column is to run an aliquot (5%) of each fraction on an agarose gel next to radioactively-labeled size standards. An autoradiogram of a gel on which cDNAs from successive column fractions has been run is shown in *Figure 6*.

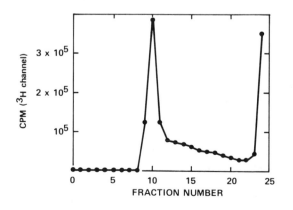

Figure 5. Elution profile of cDNAs from a Bio-Gel A-50m column. Double-stranded, *Eco*RI-linkered cDNAs were size fractionated as described in the text. Three-drop ($\sim$45 μl) fractions were collected from the column and counted on the tritium channel in a scintillation counter. (In this case, the cDNA was synthesised using a higher specific activity of 5′-[α-^{32}P]dCTP than in the text.)

9 10 11 12 13 14 15 16 17 18

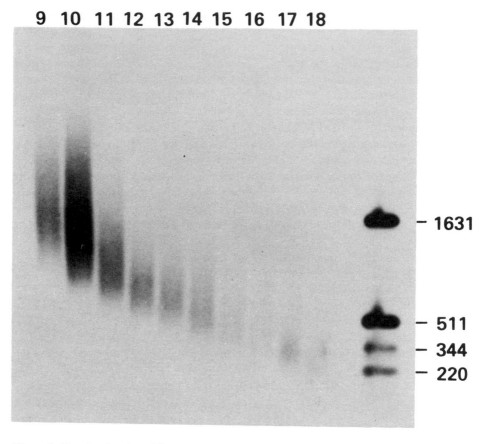

Figure 6. Size fractionation of ^{32}P-labeled cDNAs on a Bio-Gel A-50m column. See legend to *Figure 5*. Samples from column fractions number 9 to 18 were electrophoresed on a 1.2% agarose gel, which was then dried and exposed to X-ray film. DNA size standards at right are given in numbers of base pairs.

The size fractionation of the cDNAs on the Bio-Gel A-50m column is considered by many investigators to be the most laborious step of this cDNA cloning procedure. It is presented in detail here because, although laborious, it yields reliable results. An alternative method is fractionation in, and recovery from, a low melting temperature agarose gel (19). This method is more convenient than the column. However, associated with the agarose gel fractionation method is the risk of contaminating the cDNA fractions with linkers, which are present in vast molar excess and which may smear backward in the gel into the higher molecular weight fractions. Ligation of the linkers into the vector DNA results in the appearance of 'false' recombinants in the library. The risk of contamination by linkers can be decreased by taking steps to remove the linkers from the cDNA-linker mixture before running it on the gel. If the linker oligomers have been digested to completion by *Eco*RI, it should be possible to remove them from the mixture by repeated ethanol precipitation, spun-column chromatography (6), or spermine precipitation (20).

Table 8. *Eco*RI Digestion of λgt10 or λgt11 DNA.

1.	Combine in the following order: 45 μl of distilled water minus the volume of DNA and *Eco*RI, 5 μl 10 x *Eco*RI Buffer[a], 5 μg λ vector DNA in TE 7.5, 15 units of *Eco*RI to give a reaction volume of 50 μl.
2.	Incubate the mixture at 37°C for 30 min. Add 15 more units of *Eco*RI and continue incubating at 37°C for 30 min.
3.	Add 1 μl 1 M Na$_4$EDTA and heat at 70°C for 10 min.

[a]10 x *Eco*RI Buffer is 0.5 M Tris-HCl pH 7.5, 1 M NaCl, 0.1 M MgCl$_2$.

Table 9. Ligation of cDNAs with *Eco*RI-Digested λ Vector DNA.

1.	Pool the column fractions prepared as described in Section 2.5.8 that contain cDNAs of the desired size range for cloning. Centrifuge the pooled fractions for 15 min in a microcentrifuge in case there is any insoluble debris in the fractions.
2.	Add 1 μg of *Eco*RI-cut λ vector DNA to the supernatant in a 1.5 ml microcentrifuge tube. Add sodium acetate to 0.2 M and two volumes of ethanol. Allow the DNA to precipitate for 30 min in a dry ice/ethanol bath.
3.	Sediment the precipitate for 15 min in a microcentrifuge at 4°C. Wash the pellet once with 70% ethanol that has been chilled to −20°C. Centrifuge again for 10 min and then carefully remove the supernatant. Let the pellet air dry.
4.	Resuspend the pellet in 4 μl 10 mM Tris-HCl pH 7.5, 10 mM MgCl$_2$. In order to anneal the cohesive ends of λ, incubate the mixture at 42°C for 15 min.
5.	Add 0.5 μl of 10 mM ATP, 0.5 μl of 0.1 M DTT, 0.1 μl of 1.6 mg/ml T4 DNA ligase to give a reaction volume of 5 μl. Incubate the reaction mixture at 12−14°C for 2 to 16 h.

2.5.9 *EcoRI Digestion of* λgt10 *or* λgt11 *DNA*

Next, it is necessary to prepare some vector DNA to accept the double-stranded cDNA. The λgt10 or λgt11 DNA should be prepared as described in Section 2.4. Since there is batch-to-batch variation in the *in vitro* packaging efficiency of λ DNA preparations, it is important to check the packaging efficiency of any preparation of λ vector DNA before using it for cloning (see section 2.5.11). If the vector DNA packages satisfactorily *in vitro* then it can be cleaved with *Eco*RI following the protocol in *Table 8*.

To check that the λ vector DNA is sufficiently digested, package one microgram of *Eco*RI-cut DNA in a λ *in vitro* packaging reaction. The number of plaques obtained should be at least 10^3-fold lower than the number obtained from an equal mass of intact λ vector DNA.

2.5.10 *Ligation of cDNAs with EcoRI-digested* λ *Vector DNA*

The double-stranded cDNA that has been size fractionated following the procedure described in Section 2.5.8 can now be ligated to the *Eco*RI-cleaved vector DNA according to the protocol in *Table 9*.

In parallel with the ligation of the cDNAs to vector DNA, ligate one microgram of *Eco*RI-cut λ vector DNA without insert DNA, and package it *in vitro*. Titer the packaging mix on the appropriate indicator strain to determine how many false recombinant phage appear independent of added insert DNA. For λgt10, the assay consists of a clear *versus* turbid plaque test using BNN93 as host.

Directions for this assay are given in Section 2.5.12. In the case of λgt11, the assay consists of a colorless *versus* blue plaque phenotype on host Y1088 in the presence of IPTG and Xgal as described in Section 2.5.13. The background level of insert-independent false recombinant phage (for λgt10, clear plaques; for λgt11, colorless plaques) is subtracted when calculating the efficiency of cDNA cloning. The percentage of background false recombinants should not increase after the vector DNA is cut and religated. If the background increases after the vector DNA is cut and religated, contamination of the λ DNA, enzymes, or buffers by nuclease and/or DNA fragments is suspected.

2.5.11 *In Vitro Packaging of* λ *Hybrids*

In vitro packaging extracts are prepared and packaging reactions carried out according to Protocol II in reference 6, or as described in reference 5.

2.5.12 *Plating Out* λgt10 *Hybrid Phage for Screening or Amplification*

The following operations are carried out using BNN93 and BNN102 plating cells prepared according to the directions given in Section 2.4.1.

Determine what fraction of the phage is recombinant (clear plaque-forming) by plating out a small aliquot (0.01 − 0.1%) of the packaging mix using BNN93 as host. For this assay, use regular LB plates (pH 7.5), *not* plates containing 0.2% glucose, because the clear and turbid plaques are easier to distinguish on regular LB plates. It is also advisable to prepare rapid lysate DNAs from several independent clear-plaque phage as described by Davis *et al.* (11) and analyse their restriction digestion patterns to ascertain that they contain insert fragments. In λgt10, small inserts can be detected most easily if the hybrid phage DNA is cut with *Hind*III or *Bam*HI instead of *Eco*RI, because cleavage sites for these enzymes flank the site of insertion in the vector.

To screen the library with a nucleic acid probe, plate the *in vitro* packaging mix on the *hflA*150 host BNN102 on LB plates (pH 7.5) so that plaque filter replicas can be prepared as described in references 6 and 11. Use the number of *clear* plaque-forming units per ml to determine the density of plaques on the plate, since the turbid phage will form plaques at only a low efficiency on the *hflA*150 host. λgt10 is a very healthy phage and forms very large plaques. If it is necessary to decrease the plaque size in order to fit a larger number per plate for screening, this can be accomplished by (i) increasing the number of plating cells or (ii) using slightly older and dryer LB plates. After the library has been screened, it can be recovered by overlaying the plates with λ diluent.

In order to avoid creating a bias in the library against relatively poorly-growing recombinants, we recommend that the initial packaging mix of a λgt10 library be screened with the probe. However, if the library is to be amplified prior to being screened, plate the phage out at a density of 10^5 clear plaque-forming units per 150 mm Petri dish, using 125 μl BNN102 plating cells per plate and fresh LB plates (pH 7.5). Prepare plate stocks as described by Davis *et al.* (11). The titer of the plate stock is generally in the range of 5 x 10^{10} to 10^{11} p.f.u./ml. Store the plate stock at 4°C over chloroform. For long term storage, bring the plate stock to 7% dimethylsulfoxide (DMSO) and store it at $-70°C$. Such treatment reduces the titer of the library by no more than a factor of $2-3$ over a period of several years.

2.5.13 *Plating Out* λgt11 *Hybrid Phage for Amplification*

The *lon* protease-deficient host Y1090 used to plate out λgt11 libraries for screening with antibody probes is not defective for host-controlled restriction and modification enzyme activities. Therefore, the initial packaging mix of libraries cloned in λgt11 should not be plated directly on Y1090 for screening, but should first be amplified through Y1088, which is *hsdR⁻ hsdM⁺*.

Prepare Y1088 plating cells according to directions given in Section 2.4.1. To avoid creating a bias against relatively poorly-growing recombinants during amplification of the library, it is important to repress the production of toxic polypeptides (in the form of β-galactosidase fusion proteins) encoded by the recombinant phage. Therefore, *E. coli* Y1088, which produces *lac* repressor, is used as the host to amplify the library.

To determine what percentage of the phage is recombinant and to ascertain the complexity of the library, plate out a small number of phage from the packaging mix (~ 100) on *E. coli* Y1088 at $42°C$, using 2.5 ml LB soft agar (pH 7.5) containing 40 μl of 40 mg/ml Xgal and 40 μl of 1 M IPTG for a 90 mm Petri dish. Plaques produced by the parental λgt11 phage are blue on this medium, while plaques produced by recombinant phage are colorless. (Occasionally, particular recombinant phage plaques produce a small amount of blue color.) It is also advisable to prepare rapid lysate DNAs from several independent colorless-plaque phage as described by Davis *et al.* (11) and analyse their restriction digestion patterns to ascertain that they contain insert fragments.

To amplify the library, plate out the library at a density of 10^6 p.f.u. per 150 mm Petri dish, using 600 μl of Y1088 plating cells per plate and fresh LB plates (pH 7.5). Be certain to incubate the plates at $42°C$. Prepare plate stocks as described by Davis *et al.* (11). The titer of the plate stock is generally 5 x 10^{10} to 10^{11} per ml. Store the plate stock as described in Section 2.5.12.

The λgt11 library can be screened directly without prior amplification by the following procedure. For each 150 mm Petri dish, infect 0.1 ml Y1088 plating cells with $\leq 10^5$ plaque-forming units at $37°C$ for 15 minutes, then add 0.5 ml Y1090 plating cells. Immediately add 7.5 ml LB soft agar and pour onto a two-day-old LB plate (pH 7.5). Continue with step (vi) in Section 4.2 to grow up phage plaques for the preparation of nitrocellulose filter replicas.

Attempts to devise a genetic method for the selection of λgt11 recombinant phage at the expense of non-recombinant phage have been unsatisfactory because they depend upon the relative ability of cells harboring recombinant and non-recombinant phage to permit λ growth. This approach is unsatisfactory because polypeptides expressed by some recombinant phage are toxic to the host cells. For this reason, attempts to obtain high ratios of recombinant to non-recombinant phage in a λgt11 library must take place during the *construction* of the library. Two strategies are suggested here to improve the ratio of recombinants to nonrecombinants. The simplest and most successful approach is to ligate insert and vector DNAs in a tiny volume so that insert DNA is present in molar excess. One microgram of λgt11 DNA cleaved with *Eco*RI plus 0.5 microgram of insert cDNA ligated in a 2 μl volume (a smaller ligation volume than the one given in the basic procedure, Section 2.5.10 and *Table 9*) generally produces a library containing approximately 30% recombinants. One should be aware, however, that occasionally multiple foreign DNA fragments are inserted into a single vector DNA molecule when the ratio of insert to vector DNA is this high. The second approach is to treat the vector DNA with alkaline phosphatase, removing the 5' terminal phosphate required for ligation. Thus, the two vector arms cannot be ligated together in the absence of a foreign DNA insert fragment. Unfortunately, the quality of most commercially available alkaline phosphatase is such that the viability of vector DNA subsequent to treatment is often greatly reduced. For this reason, if alkaline phosphatase is used, the enzyme should be titrated to determine the minimum amount required to reduce the ability of *Eco*RI-digested vector DNA to be ligated (in the absence of insert DNA) by 90−95%.

2.6 Examples of Results

2.6.1 *Efficiency of cDNA Cloning*

The efficiencies of cDNA cloning that various individuals have obtained using the cDNA cloning procedure presented above and λgt10 as a vector are presented in *Table 10*. Similar data for the vector λgt11 appear in *Table 11*. The average of the efficiencies listed in the two tables is 6.6×10^5 recombinants per microgram

Table 10. Efficiency of cDNA Cloning in λgt10.

Poly(A)$^+$ RNA source	No. clones/μg poly(A)$^+$ RNA
Corn seedling[a]	2.5×10^5
Pea shoot segments[b]	3.0×10^5
Yeast[c]	7.5×10^5
Nematode[d]	5.0×10^5
Bovine retina[e]	1.3×10^5

[a]T.Huynh, unpublished.
[b]A.Theologis, personal communication.
[c]R.Young, unpublished.
[d]B.Meyer, personal communication.
[e]J.Nathans, personal communication.

Table 11. Efficiency of cDNA Cloning in λgt11.

Poly(A)$^+$ RNA source	No. clones/μg poly(A)$^+$ RNA	% of total p.f.u. in library
Yeast[a]	1.5×10^6	15%
Nematode[b]	1.2×10^6	8%

[a]R.Young unpublished.
[b]B.Meyer, personal communication.

of polyadenylated RNA. These results demonstrate that it is feasible to construct a cDNA library containing 10^5 to 10^7 recombinants in a λ vector starting with a few micrograms of poly(A)$^+$ RNA. Such a library should be large enough to contain sequences representing even the rarest mRNAs.

The efficiency of cDNA cloning depends heavily upon the quality of the reagents used. Not the least important of these is the poly(A)$^+$ RNA template. Before being used for the synthesis of cDNA, the poly(A)$^+$ RNA should be checked rigorously for signs of degradation. The availability of high quality, concentrated enzyme stocks free from contamination with nuclease and phosphatase activities is also essential. Many investigators prefer homemade enzyme preparations to commercial sources. Another important factor is the efficiency of the *in vitro* packaging extracts. The cloning efficiencies listed in *Tables 10* and *11* are based on reactions performed using packaging extracts ranging in efficiency from 5 x 10^7 to 10^8 plaque-forming units per microgram of vector DNA. An advantage of the dependency of the cDNA cloning efficiency on the quality of the packaging extracts is that the cDNA cloning procedure will become more efficient as better methods of preparing *in vitro* packaging extracts are discovered.

2.6.2 *Size of cDNA Clones*

To ascertain that the phage did indeed contain insert fragments, DNA was prepared from twenty-four randomly chosen independent clear-plaque phage from the λgt10 corn cDNA library listed in *Table 10*, and digested with *Eco*RI. All twenty-four contained *Eco*RI fragment inserts, with an average length of 650 base pairs.

From the average size of cDNA clone prepared by this method, it is apparent that the average clone is less than full-length. To a large extent, the average size of cDNA clone depends upon what size fractions of cDNA are included in the final ligation reaction. Many investigators, using either nucleic acid or antibody probes, consider a library of predominantly partial cDNA clones compatible with their strategy for isolating a particular gene. Once a particular cDNA clone has been isolated, it in turn is used as a probe to screen the library for longer cDNA clones. Using this stepwise approach, it is usually possible to isolate a cDNA clone containing the entire coding region of an mRNA from a large cDNA library.

2.6.3 *Structural Features of cDNA Clones*

The procedure for the preparation of double-stranded cDNA presented in this chapter includes a widely-used step involving nuclease S1 (12). This approach to the preparation of double-stranded cDNA relies on self-priming of second strand cDNA synthesis by hairpin structures formed at the 3′ termini of the first strand cDNAs. Certain structural anomalies associated with this approach have been noted by other investigators and will be mentioned here. First, because nuclease S1 is used to digest the hairpin structure, the cDNA clones inevitably lack sequences corresponding to the 5′ ends of the mRNAs. In spite of this deficiency, recovering cDNA clones containing sequences upstream of the ATG defining the start site of translation from large cDNA libraries is generally not a problem. A second structural anomaly associated with the use of nuclease S1 is that occasionally a cDNA clone is found containing sequences at its 5′ terminus which are not colinear with the mRNA (21). The cause of this artifact appears to be incomplete digestion by nucleae S1. Since this artifact is associated with the use of nuclease S1 during double-stranded cDNA synthesis, it is important to ascertain that any particular cDNA clone under study is colinear with the mRNA along its entire length. This is done by performing S1 mapping experiments as described in reference 6.

Both structural anomalies associated with the 5′ ends of the cDNA clones described above could be avoided by a method of cDNA preparation designed to enrich for full-length cDNA clones. There is no reason that methods avoiding the use of nuclease S1 (21,22) could not be adapted for use with λgt10 and λgt11.

3. SCREENING cDNA LIBRARIES IN λgt10 AND λgt11 WITH NUCLEIC ACID PROBES

Recombinant DNA libraries in λ vectors are screened with nucleic acid probes by hybridising the probes with replicas of plaques on nitrocellulose filters. Procedures for screening cDNA libraries in λ vectors are identical to procedures for screening genomic DNA libraries. Much of this methodology is well-developed and is presented in other methods manuals (6,11). Protocols for preparing nitrocellulose filter plaque replicas and for carrying out hybridisation reactions can be found in references 6 and 11. Generally, following a prehybridisation wash, the nitrocellulose filter replicas are incubated with the ^{32}P-labeled denatured or single-stranded DNA probe in a hybridisation buffer, sometimes containing formamide, at a temperature approximately 25°C below the melting temperature of the duplex expected between the probe and the DNA fixed to the nitrocellulose filter. The hybridisation buffer may contain the polymer dextran sulfate to enhance the rate and extent of hybridisation (23). After hybridisation is complete, the filters are washed several times at a level of stringency appropriate for the experiment. The filters are then blotted dry and exposed to X-ray film.

Depending on the investigator's strategy to isolate a particular cDNA clone, the nucleic acid probe may take one of several forms.

(i) If the probe is a previously-cloned DNA fragment, usually the plasmid containing the fragment is labeled by 'nick translation', a method in which *E. coli* DNA polymerase I is used *in vitro* to replace nucleotides on the DNA fragment with radioactive nucleotides. Protocols for performing nick translation reactions can be found in references 6 and 11.

(ii) The probe may consist of a mixture of synthetic deoxyoligonucleotides predicted from a particular amino acid sequence. If the oligonucleotides are sufficiently long, they can be 5′-end labeled and used directly as probes to screen the cDNA library under carefully controlled hybridisation conditions (24). Alternatively, as long as the oligonucleotides represent the DNA strand complementary to the messenger RNA, the oligonucleotides can be annealed individually to unfractionated poly(A)$^+$ RNA under carefully defined conditions and extended with reverse transcriptase to identify the one which stimulates the synthesis of the correct DNA sequence (25).

Once the correct primer has been identified, it is used to synthesise cDNA probe using reverse transcriptase in the presence of radioactive nucleotides and poly(A)$^+$ RNA. This produces a highly specific, single-stranded probe for screening the cDNA library.

(iii)   The probe may consist of RNA or single-stranded cDNA made by reverse transcribing poly(A)$^+$ RNA. A method for preparing kinase-labeled RNA appears in reference 26. A method for preparing cDNA probe can be found in reference 27. When either RNA or cDNA is used as a probe, it is important to include polyuridylic acid (Pharmacia P-L Biochemicals, Inc., or Collaborative Research Inc.) in the prehybridisation and hybridisation buffers at a concentration of 250 μg/ml. The polyuridylic acid block is necessary to block hybridisation of the probe to regions in the cDNA clones corresponding to the poly(A) tails at the 3' ends of the mRNAs.

4. SCREENING LIBRARIES IN λgt11 WITH ANTIBODY PROBES

λgt11 libraries are screened with antibody probes as plaques on a lawn of *E. coli* Y1090. Y1090 contains:

(i)   *lac* repressor, which prevents *lacZ*-directed gene expression until it is derepressed by the addition of IPTG to the medium;

(ii)   A deficiency in the *lon* protease, which increases the stability of the recombinant fusion protein;

(iii)   *supF* to suppress the phage mutation causing defective lysis (*S100*).

To ensure that fusion proteins toxic to the host produced by particular recombinants will not inhibit the growth of particular members of the library, plaque formation is initiated without expression from the *lacZ* gene promoter. After the number of infected cells surrounding the plaques is large, *lacZ*-directed gene expression is switched on by the addition of IPTG. The IPTG is added by placing an IPTG-impregnated nitrocellulose filter disk over the lawn. After a few more hours of incubation, the nitrocellulose filter is removed from the lawn with protein from lysed cells in the plaques bound to it. The filter is washed, incubated with the antibody solution, and then with ^{125}I-labeled *Staphylococcus aureus* protein A to visualise antibody binding.

4.1 Reagents

(i)   Nitrocellulose filter circles, 0.45 μm pore size, diameter 82 mm or 132 mm: Schleicher and Schuell.

(ii)   Isopropyl β-D-thiogalactopyranoside (IPTG): Sigma Chemical Company. Prepare a 1 M solution in sterile distilled water. Store at $-20\,^\circ$C.

(iii)   Fetal calf serum: HyClone. Filter before use through nitrocellulose, 0.45 μm pore size.

(iv)   *Staphylococcus aureus* ^{125}I-protein A, specific activity 30 mCi/mg: ICN.

4.2 Procedure for Screening Libraries in λgt11 With Antibody Probes

(i)   Streak out *E. coli* Y1090 to give single colonies on LB plates (pH 7.5) containing 50 μg/ml ampicillin. Incubate at 37°C.

(ii)   Starting with a single colony, grow Y1090 to saturation in LB (pH 7.5) at 37°C with good aeration.

(iii) If the plating is to be done on 90 mm plates, mix 0.2 ml of the Y1090 culture with 0.1 ml of λ diluent containing 3 x 10⁴ p.f.u. of the λgt11 library for each plate. If the plating is to be done on 150 mm plates, use 0.6 ml of the Y1090 culture with up to 10⁵ p.f.u. in λ diluent for each plate.

(iv)   Adsorb the phage to the cells at 37°C for 15 min.

(v)   Add 2.5 ml (for a 90 mm plate) or 7.5 ml (for a 150 mm plate) of LB soft agar (pH 7.5) to the culture and pour onto an LB plate (pH 7.5). Use dry plates (i.e. two days old) so that the soft agar will stick to the plate.

(vi)   Incubate the plates at 42°C for three to four hours.

(vii) Remove the plates to a 37°C incubator. Do not permit the plates to cool below 37°C.

(viii) Overlay each plate with a dry nitrocellulose filter disk which has been saturated previously in 10 mM IPTG in water.

(ix)   Incubate for 2 – 3 h longer at 37°C. (A second filter may be overlaid after the first has been removed and incubated for an additional 2 – 3 h if a duplicate is required.)

(x)   Remove the plates to room temperature, quickly mark the position of the filter on the plate with a needle dipped in waterproof ink, and remove the filters carefully.

(xi)   Do not allow the filters to dry out during any of the subsequent steps. Perform all of the following washing and incubation steps at room temperature with gentle shaking. First, rinse the filters briefly in TBS. (TBS is 50 mM Tris-HCl pH 8.0, 150 mM NaCl.)

(xii) Incubate the filters in TBS + 20% fetal calf serum for 15 – 30 min. Use 5 ml per 82 mm filter and 10 ml per 132 mm filter.

(xiii) Incubate the filters in TBS + 20% fetal calf serum + antibody for 1 h. Use 5 ml per 82 mm filter and 10 ml per 132 mm filter. Dilute serum antibody or purified IgG (at 10 mg/ml) one hundred-fold in TBS + 20% fetal calf serum. (This solution can be reused several times.)

(xiv) Wash the filters in TBS for 5 – 10 min.

(xv)   Wash the filters in TBS + 0.1% NP-40 for 5 – 10 min.

(xvi) Wash the filters again in TBS alone for 5 – 10 min.

(xvii) Rinse the filters briefly in TBS + 20% fetal calf serum, using 5 ml per 82 mm filter and 10 ml per 132 mm filter.

(xviii)Transfer the filters to TBS + 20% fetal calf serum contining ¹²⁵I-protein A (10⁶ c.p.m./82 mm filter, specific activity 30 mCi/mg). Use 5 ml per 82 mm filter and 10 ml per 132 mm filter. Incubate for 1 h at room temperature.

(xix) Wash the filters in TBS, TBS + 0.1% NP-40, then in TBS as in steps (xiv) – (xvi).

(xx)   Dry the filters and expose to Kodak X-Omat AR film with a Cronex Lightning Plug intensifying screen at −70°C.

(xxi) To retest a putative positive signal, remove an agar plus containing phage particles from the region of the plate corresponding to the signal on the filter. Incubate the agar plug in 1 ml of λ diluent for at least one hour.

(xxii) Replate the phage and repeat the screening procedure with the antibody probe until all the plaques on the plate produce a signal.

4.3 Further Comments on the Procedure for Screening Libraries with Antibody Probes

(i) Polyvalent antibodies have been used successfully to isolate genes using λgt11 libraries constructed with cDNA or genomic DNA from organisms with genomes ranging in complexity from bacteria to mammals. The quality of the antibody probe is clearly important. Polyvalent antibodies may have an advantage over monoclonal antibodies when used as probes, because the multi-clonal population is rarely restricted to the recognition of a single epitope. As expected, high titer antibodies produce better signals than low titer antibodies. It is reasonable to assume that antibodies which produce good signals on a 'Western' blot will produce good signals in the λgt11 screening procedure.

(ii) Polyvalent antibodies often contain components which bind to antigens normally produced by *E. coli*. In order to avoid the high background of signals produced by this binding activity, steps must be taken to remove it from the antibody preparation. This can be accomplished most effectively by immobilising an *E. coli* lysate on a Sepharose resin (using CnBr-activated Sepharose, for example), or on nitrocellulose filters, and incubating the antibody with the bound bacterial lysate. Prepare the *E. coli* lysate from BNN97 by following the procedure presented in Section 5. Sonicate the lysate or treat it with DNase to reduce its viscosity. Bind the lysate to CnBr-activated Sepharose following the manufacturer's protocol. To bind the lysate to nitrocellulose filters, simply dip the filters in the lysate, then wash the filters in TBS. Next, dilute the antibody approximately ten-fold in TBS + 20% fetal calf serum, and incubate it with the Sepharose resin or the nitrocellulose filters for 15 – 30 min. Remove the antibody from the solid substrate (centrifugation suffices to remove the Sepharose beads), and incubate it with a second batch of resin or filters. Two independent washes are usually sufficient to purify away most of the anti-*E. coli* component in the antibody. Because the antibody is reused during the screening process to retest putative positive signals, anti-*E. coli* components are removed to a greater extent the more the probe is used, and backgrounds often progressively decrease.

(iii) The diluted antibody preparation can be used repeatedly. Sometimes as many as ten separate screens can be performed with the same diluted antibody preparation.

(iv) While fetal calf serum is an excellent protective agent for antibody, one significantly less expensive alternative is to use 10 mg/ml chicken egg albumin (Sigma Chemical Company). However, it is not yet clear whether the latter performs as well as fetal calf serum when a single diluted antibody preparation is used for multiple screens.

5. PREPARATION OF RECOMBINANT ANTIGEN FROM A λgt11 RECOMBINANT LYSOGEN

It is often useful to have preparative amounts of a polypeptide specified by a cloned piece of DNA. For some purposes (for instance, radioimmunoassays), it is sufficient to have a crude *E. coli* lysate containing an antigen specified by the cloned DNA of interest. A crude lysate containing a particular recombinant antigen can be prepared easily by expressing a λgt11 recombinant as a lysogen in *E. coli* Y1089. Y1089 contains:

(i)   the *lac* repressor (*lacI* gene product), which prevents *lacZ*-directed gene expression until derepressed by the addition of IPTG to the medium;

(ii)   a deficiency in the *lon* protease, which increases the stability of the recombinant fusion protein;

(iii)   a mutation which enhances the frequency of phage lysogeny (*hflA*150).

To produce the recombinant fusion protein, Y1089 is lysogenised with the λgt11 clone of interest. The lysogen is grown to high cell density, *lacZ*-directed fusion protein production is induced by the addition of IPTG to the medium, and the cells are harvested and lysed.

5.1 Generation of a λgt11 Recombinant Lysogen in Y1089

(i)   Grow Y1089 cells in LB medium (pH 7.5)/0.2% maltose at 37°C to saturation.

(ii)   Infect the Y1089 cells with the λgt11 recombinant phage at a multiplicity of approximately 5 for 20 min at 32°C in LB medium (pH 7.5) supplemented with 10 mM $MgCl_2$.

(iii)   Plate the cells on LB plate (pH 7.5) at a density of approximately 200 per plate and incubate at 32°C. (At this temperature, the temperature-sensitive phage repressor is functional.)

(iv)   Test single colonies for temperature sensitivity at 42°C. Spot cells from single colonies using sterile toothpicks onto two LB plates (pH 7.5). Incubate the first plate at 42°C and the second at 32°C.

(v)   Clones which grow at 32°C but not at 42°C are assumed to be lysogens. Lysogens arise at a frequency between 10% and 70%.

5.2 Preparation of a Crude Lysate From a λgt11 Recombinant Lysogen

(i)   Inoculate 100 ml of LB medium (pH 7.5) with a single colony of the Y1089 recombinant lysogen. Incubate the culture at 32°C with good aeration.

(ii)   When the culture has grown to an optical density of 0.5 measured at 600 nm, increase the temperature of the culture to 42−45°C as rapidly as possible, and incubate the culture at the elevated temperature for 20 min with good aeration.

(iii)   Add IPTG to 10 mM.

(iv)   Incubate the culture at 37−38°C for approximately 60 min. (Do not let the temperature of the culture drop below 37°C.) At this stage, the Y1089 lysogen will sometimes lyse, even though Y1089 does not suppress the mutation causing defective lysis (*S100*) in λgt11. This is a consequence firstly of the 'leakiness' of the *S100* amber mutation and secondly because the accumulation of foreign proteins in *E. coli* often renders it susceptible to lysis. For this reason, the longest incubation time achievable at $37-38°C$ without lysis occurring should be determined for each individual recombinant lysogen.

(v)   Harvest the cells as rapidly as possible in a Beckman JA-10 rotor at 5000 r.p.m. for 5 min at $24-37°C$. (A sudden shift in temperature during harvest appears to increase the rate of proteolysis in experiments done on a small number of different lysogens. Therefore, centrifugation is done at a temperature between 24 and 37°C.)

(vi)   Rapidly resuspend the cells in 1/20 to 1/50 of the original culture volume in a buffer suitable for proteins.

(vii)   Immediately freeze the resuspended cells in liquid nitrogen.

(viii) Thawing of the frozen cells results in essentially complete lysis of the induced lysogen.

5.3 Preparation of β-Galactosidase Fusion Protein From a Crude Lysate of a λgt11 Recombinant Lysogen

If crude antigen is required, then the crude lysate described in Section 5.2 can be used directly. If pure antigen is needed, the β-galactosidase fusion protein can be purified in several different ways. The most rapid method of purification takes advantage of the size of the β-galactosidase fusion protein. (The β-galactosidase portion of the fusion accounts for 114 kilodaltons.) Since only a few proteins in *E. coli* are larger than β-galactosidase, the fusion protein is often resolved from other proteins on SDS-polyacrylamide gels. Preparative gels can be used to isolate large quantities of denatured protein. If pure antigen in native form is required, then the fusion protein is purified by classical column chromatography.

6. ACKNOWLEDGEMENTS

We gratefully acknowledge the contributions of many people to these methods. We are indebted to Stewart Scherer who pointed out to us the need for the development of λ vectors suitable for cDNA cloning and expression. The initial steps in the cDNA synthesis procedure represent a revision of a protocol by David Kemp. To Stewart Scherer and Barbara Meyer we owe the suggestion that an *hfl* host could be used to select against cI^+imm^{434} phage. Thomas St.John contributed to discussions about both the cDNA synthesis procedure and the development of methods to screen plaque replicas with antibody probes. We thank Andrew Hoyt who generously provided the *hflA*150 mutant, and Jeremy Nathans who provided helpful suggestions as well as the autoradiograms which appear in *Figures 4* and *6*. We also wish to thank the many individuals who tested the procedures and offered critical suggestions. This work was supported by grants to R.W.D. from BARD and NIH 6M 21891.

7. REFERENCES

1. Young,R.A. and Davis,R.W. (1983) *Proc. Natl. Acad. Sci. USA*, **80**, 1194.
2. Young,R.A. and Davis,R.W. (1983) *Science*, **222**, 778.
3. Murray,N.E., Brammar,W.J. and Murray,K. (1977) *Mol. Gen. Genet.*, **150**, 53.
4. Murray,N.E. (1983) in *Lambda II*, Hendrix,R.W., Roberts,J.W., Stahl,F.W. and Weisberg,R.A. (eds.), Cold Spring Harbor Laboratory, Cold Spring Harbor, p. 395.
5. Scherer,G., Telford,J., Baldari,L. and Pirrotta,V. (1981) *Dev. Biol.*, **86**, 438.
6. Maniatis,T., Fritsch,E.F. and Sambrook,J. (1982) *Molecular Cloning*, Cold Spring Harbor Laboratory, Cold Spring Harbor, New York.
7. Hoyt,M.A., Knight,D.M., Das,A., Miller,H.I. and Echols,H. (1982) *Cell*, **31**, 565.
8. Goldberg,A.L. and St. John,A.C. (1976) *Annu. Rev. Biochem.*, **45**, 747.
9. Charnay,P., Gervais,M., Louise,A., Galibert,F. and Tiollais,P. (1980) *Nature*, **286**, 893.
10. Kemp,D.J., Coppel,R.L., Cowman,A.F., Saint,R.B., Brown,G.V. and Anders,R.F. (1983) *Proc. Natl. Acad. Sci. USA*, **80**, 3787.
11. Davis,R.W., Botstein,D. and Roth,J.R. (1980) *Advanced Bacterial Genetics*, Cold Spring Harbor Laboratory, Cold Spring Harbor, New York.
12. Efstratiadis,A., Kafatos,F.C., Maxam,A.M. and Maniatis,T. (1976) *Cell*, **7**, 279.
13. Wickens,M.P., Buell,G.N. and Schimke,R.T. (1978) *J. Biol. Chem.*, **253**, 2483.
14. Seeburg,P.H., Shine,J., Martial,J.A., Baxter,J.D. and Goodman,H.M. (1977) *Nature*, **270**, 486.
15. Modrich,P. and Zabel,D. (1976) *J. Biol. Chem.*, **251**, 5866.
16. Rubin,R.A. and Modrich,P. (1977) *J. Biol. Chem.*, **252**, 7265.
17. Murray,K. and Murray,N.E. (1975) *J. Mol. Biol.*, **98**, 551.
18. Wahl,G.M., Padgett,R.A. and Stark,G.R. (1979) *J. Biol. Chem.*, **254**, 8679.
19. Sealey,P.G. and Southern,E.M. (1983) in *Gel Electrophoresis of Nucleic Acids: A Practical Approach*, Rickwood,D. and Hames,B.D. (eds.), IRL Press, Oxford, p. 39.
20. Hoopes,B.C. and McClure,W.R. (1981) *Nucleic Acids Res.*, **9**, 5493.
21. Land,H., Grez,M., Hauser,H., Lindenmaier,W. and Schütz,G. (1981) *Nucleic Acids Res.*, **9**, 2251.
22. Gubler,U. and Hoffman,B.J. (1983) *Gene*, **25**, 263.
23. Wahl,G.M., Stern,M. and Stark,G.R. (1979) *Proc. Natl. Acad. Sci. USA*, **76**, 3683.
24. Suggs,S.V., Wallace,R.B., Hirose,T., Kawashima,E.H. and Itakura,K. (1981) *Proc. Natl. Acad. Sci. USA*, **78**, 6613.
25. Houghton,M., Stewart,A.G., Doel,S.M., Emtage,J.S., Eaton,M.A.W., Smith,J.C., Patel,T.P., Lewis,H.M., Porter,A.G., Birch,J.R., Cartwright,T. and Carey,N.H. (1980) *Nucleic Acids Res.*, **8**, 1913.
26. Maizels,N. (1976) *Cell*, **9**, 431.
27. St.John,T.P. and Davis,R.W. (1979) *Cell*, **16**, 443.

CHAPTER 3

An Alternative Procedure for the Synthesis of Double-stranded cDNA for Cloning in Phage and Plasmid Vectors

CHRISTINE J.WATSON and JAMES F.JACKSON

1. INTRODUCTION

The main purpose of this chapter is to describe an alternative protocol for synthesis of double-stranded cDNA suitable for molecular cloning in a variety of vectors. It is based on a novel method for second strand cDNA synthesis introduced by Okayama and Berg (1) in the context of a plasmid cDNA cloning protocol. Classically, the mRNA-cDNA hybrid that is the product of the first strand reaction is treated with alkali to hydrolyse the RNA. Second strand synthesis is primed from a 'hairpin' structure found at the 3' end of cDNA molecules (2). This structure, as well as any nucleotides from the 5' end of the first strand not copied during the second strand reaction (and therefore single-stranded) are prepared for linker addition by treatment with S1 nuclease from *Aspergillus oryzae*. In our protocol, the mRNA-cDNA hybrid that is the product of the first strand reaction is not destroyed by alkali, but is used as a substrate for treatment with RNase H, *E. coli* DNA polymerase I and *E. coli* DNA ligase in the presence of deoxyribonucleoside triphosphates. RNase H introduces nicks in the RNA to provide Okazaki fragment-like primers, which provide a substrate for the replacement of the RNA by DNA (mediated by the pol I). The *E. coli* DNA ligase will seal the nicks, but unlike T4 DNA ligase, cannot ligate resultant double-stranded DNA molecules to each other (which would generate artefactual cDNAs). The double-stranded cDNA, after linker addition and digestion to generate cohesive ends, can be cloned into vectors specifically designed for screening with either nucleic acid or antibody probes. The method described in detail below is that likely to be of most general interest − preparing *Eco*RI-terminated double-stranded cDNA to be cloned into high efficiency bacteriophage lambda vectors. An independent report describing a similar protocol has recently appeared (3).

Immunity insertion vectors like NM607 (8), NM1149 (4), λgt10 (see Chapter 2 of this book) and Charon 7 or Charon 16 (5) offer cloning sites for *Eco*RI-terminated molecules. Charon 7 and NM1149 can also be used to clone *Hind*III-terminated molecules in similar fashion. Libraries prepared in these vectors are normally screened by hybridisation with nucleic acid probes. They offer cloning efficiencies exceeding 2×10^7 plaque forming units (p.f.u.)/μg double-stranded cDNA both in our hands and in those of others (6). The high efficiency is due, in our opinion, to three major factors:

(i)   *In vitro* packaging protocols permit the recovery of lambda phage genomes at an efficiency of 1%, much higher than those achieved by transformation of plasmids.

(ii) Since non-recombinant (i.e., reconstituted vector) molecules can be eliminated in these vectors by selection on NM514, a restriction-deficient derivative of POP*lyc*7 (7) or its equivalent, *Hfl*A (Huynh, Young and Davis, Chapter 2), dephosphorylation of vector DNA is *not* necessary. Ligation of double-stranded cDNA to vector can be performed in a molar excess of vector molecules, which ensures most cDNA molecules will be ligated to a vector molecule at both ends, a requirement for packaging *in vitro*. For a detailed explanation of these vectors, see reference 8 and Chapter 2 of this volume.

(iii) Yields of double-stranded cDNA prepared by our alternative protocol appear to be quite good. The yield of second strand product is normally 100%. Additionally, losses due to nuclease S1 treatment are avoided.

The protocol is also suitable for preparing cDNA for cloning in 'expression vectors' of phage (see Chapter 2) or plasmid (9,10) origin which can be screened with antibody probes. In these cases, it is normally necessary to dephosphorylate the vector to reduce contamination of recombinants with reconstituted vector molecules.

2. SYNTHESIS OF cDNA

The procedures described here can be used as alternatives to those described in Sections 2.5.1 to 2.5.8 of the previous chapter. The precautions described there for carrying out the reactions apply equally well to this set of protocols. It is extremely important that siliconised microcentrifuge tubes be used for all reactions. Tubes may easily be siliconised by placing a beaker of tubes and a beaker with a small amount of siliconising compound in a desiccator under vacuum for several hours. The authors use dichlorodimethylsilane, which is supplied by BDH as a 2% solution in 1,1,1-trichloroethane. The tubes are then rinsed in sterile distilled water and autoclaved.

2.1 First Strand cDNA Synthesis

The quality of first strand transcripts is dependent upon several factors, some of which are discussed in Chapter 2. As the authors point out, use of undegraded mRNA is important to produce a high quality cDNA library. Procedures involving lysis in guanidinium thiocyanate (11,12) give good results with a variety of tissues, as assayed by Northern blots using either the probe which will be used to screen the library or a clone coding for a ubiquitous, conserved sequence such as a ribosomal protein, actin, tubulin or histone.

The protocol for the first strand reaction is given in *Table 1*. The reaction conditions are very similar to those given in Chapter 2. The addition of the ribonuclease inhibitor RNasin has been found to be particularly effective in im-

Table 1. First Strand cDNA Synthesis.

1. For a 100 μl reaction, heat 10 μg of polyadenylated RNA in 10 μl of water for 5 min at 65°C. Chill the tube on ice and transfer the RNA to a tube containing 5 μl of actinomycin D stock solution[a] which has been left in the air to allow all the solvent to evaporate.

2. Add the following: 20 μl of 5 x RT buffer[b], 20 μl of 5 x d(ACGT)TPs[c], 10 μl of 200 μg/ml oligo(dT)$_{12-18}$ (Collaborative Research), 1 μl of 1 M DTT[d], 60 units of RNasin[e], 100 units of reverse transcriptase[f], 10 μCi α[^{32}P]dGTP[g] and water to bring the reaction volume to 100 μl.

3. Take two 2 μl aliquots to determine the 'total' and 'zero-time' radioactivity as described in the text, and then incubate the reaction tube at 42°C for 2 h.

4. Take a 2 μl aliquot to determine the radioactivity incorporated into DNA and then add 5 μl of 0.5 M EDTA pH 8.0, 2.5 μl of 10% SDS, 50 μl of phenol (equilibrated with 100 mM Tris-HCl pH 7.4), and 50 μl of chloroform. Shake to emulsify and then separate the phases by centrifugation in a microcentrifuge. Re-extract the organic phase with 50 μl of TE[h]. Ether extract the pooled aqueous phases.

5. To the aqueous phase from the ether extraction, add 100 μl of 4 M ammonium acetate (filtered through a nitrocellulose membrane), 400 μl of 95% ethanol, mix and chill in a dry ice-ethanol bath for 15 min.

6. Pellet the precipitated nucleic acid by centrifugation in a microcentrifuge. Drain the tube well and resuspend the pellet in 50 μl of TE. Add 50 μl of 4 M ammonium acetate, 200 μl of 95% EtOH.

7. Mix, chill, and centrifuge as in Step 6. Drain the tube and dry the pellet under vacuum. Resuspend the pellet in 30 μl of TE.

8. Remove a 2 μl aliquot to assay the length of the cDNA by gel electrophoresis.

[a]The stock solution of actinomycin D (Calbiochem) is made up at 800 μg/ml in 80% ethanol.
[b]5 x RT buffer is 250 mM Tris-HCl pH 8.3 at 42°C, 40 mM MgCl$_2$, 250 mM KCl.
[c]5 x d(ACGT)TP is a stock solution containing each deoxynucleoside triphosphate (PL) at a concentration of 5 mM. We prepare stocks of each triphosphate as a 20 mM solution in 10 mM Tris HCl pH 7.5.
[d]The 1 M DTT solution is filtered through a nitrocellulose membrane before use.
[e]RNasin is supplied by Biotech at 30 U/μl. One unit will inhibit by 50% the activity of 5 ng of ribonuclease A.
[f]The authors have successfully used avian myeloblastosis virus reverse transcriptase supplied by J.W.Beard. One unit of reverse transcriptase will incorporate 1 nmol of dTMP into an acid-insoluble product in 10 min at 37°C.
[g]The specific activity of the α[^{32}P]dGTP (Amersham) is 410 Ci/mmol.
[h]TE is 10 mm Tris-HCl pH 8.0, 1 mM EDTA.

proving the length of first-strand transcripts. It is extracted from mammalian liver or placenta and is commercially available from several suppliers.

This step and subsequent synthesis of the second strand can be followed by assaying the incorporation of radiolabelled nucleotide into precipitable nucleic acid. This may be carried out by spotting two 2 μl aliquots of the reaction mixture onto Whatman DE81 paper circles immediately after adding all components and before starting the incubation. One is to determine the 'total counts', and the other for the 'zero-time' background. After incubating the cDNA reaction for 2 h at 42°C (Step 3 of *Table 1*), remove 2 μl and spot it onto DE81 to determine the incorporation into 'precipitable' DNA. Wash the two filters carrying the aliquots taken at zero time and at 2 h as follows:

(i)   5 x 5 min washes in 0.5 M Na$_2$HPO$_4$.
(ii) 2 x 1 min washes in water.
(iii) 2 x 1 min washes in 100% ethanol.

This is carried out by swirling the filters in a beaker containing about 25 ml of the appropriate washing solution. Determine the fraction of [^{32}P]nucleotide that has been incorporated into single-stranded cDNA by counting the radiation from the filters, using scintillant and the ^{32}P channel of a scintillation counter. In this reaction, we typically observe 10 − 50% yields by weight of first strand product. The actual yield of cDNA is likely to be somewhat higher, as we use poly(A)-RNA purified by a single cycle of oligo(dT)-cellulose chromatography, which is 30 − 50% pure.

2.2 Second Strand cDNA Synthesis

Double-stranded cDNA made using this protocol should be identical in length to its single-stranded cDNA template when assayed by electrophoresis on an alkaline agarose gel. This is a major difference from the expectation using the method described in the previous chapter in which the denatured double-stranded cDNA is twice as long as the template, as a result of the hairpin structure that joins the two strands. In the reaction described here, the second strand is made ten times hotter than the first strand by using a 10-fold lower concentration of unlabelled nucleotides and the same amount of label. The radiolabel in the double-stranded cDNA then reflects the amount of second strand synthesis, *Figure 1*. Yields of second strand cDNA are 100 ± 20%, with the variation chiefly due to error in measurement of the radioactive tracer.

The reaction conditions are given in *Table 2*. The reaction differs considerably

Table 2. Second Strand cDNA Synthesis.

---

1. To 28 μl of the product of the first strand cDNA reaction add 25.0 μl of 4 x PolI buffer[a], 2.5 μl of 5 x d(ACGT)TP[b], 1.0 μl of 15 mM β-NAD (Sigma), 10 μCi of α[^{32}P]dGTP. Add 30 units of *E. coli* DNA polymerase I[d] and 1 unit of RNase H[e] per μg of cDNA: then 5 units of *E. coli* DNA ligase. Adjust the volume of the mixture with water so that it will be 100 μl when the three enzymes have been added.

2. Remove two 2 μl aliquots to determine the 'total' and 'zero-time' radioactivity as described in Section 2.1. Incubate the reaction for 1 h at 14°C, and then for 1 h at room temperature.

3. Remove a 2 μl aliquot to determine the radioactivity incorporated into DNA and then add 5 μl of 0.5 M EDTA pH 8.0, 2 μl of 10% SDS, 50 μl of phenol equilibrated with 100 mM Tris-HCl, pH 8.0, and 50 μl of chloroform. Shake the tube to emulsify the mixture and centrifuge in a microcentrifuge for 1 min. Re-extract the phenol phase with 50 μl of TE.

4. To the combined aqueous phases add 150 μl of 4 M ammonium acetate and 600 μl of ethanol. Chill to − 70°C in a dry ice-ethanol bath. Proceed as in steps 6 and 7 in *Table 1*. Alternatively one can precipitate the cDNA by adding 1/10 volume 3 M sodium acetate (filtered through a nitrocellulose membrane) and 2 volumes of ethanol. This may give quantitatively better yields.

---

[a]To make 1 ml of 4 x PolI buffer, mix 400 μl of 1 M Hepes pH 7.6, 16 μl of 1 M MgCl$_2$, 4.4μl of 2-mercaptoethanol, 270 μl of 1 M KCl, and 310 μl of water.
[b]See footnote [c] to *Table 1* for description of d(ACGT)TP.
[c]1 unit of *E. coli* DNA ligase is the amount of enzyme required to give 50% ligation of *Hind*III digested lambda DNA in 30 min at 16°C in 20 μl at a 5′ termini concentration of 0.12 mM (∼330 μg/ml). The authors use ligase purchased from New England Biolabs.
[d]1 unit of *E. coli* DNA polymerase is the amount of enzyme required to convert 10 nmol of total deoxyribonucleotides to an acid insoluble form in 30 min at 37°C. The authors use enzyme purchased from New England Biolabs.
[e]1 unit of RNase H hydrolyses 1 nmol of [^{3}H]poly(A) in [^{3}H]poly(A):poly(dT) in 20 min at 37°C. The authors use RNase H purchased from Bethesda Research Laboratories.

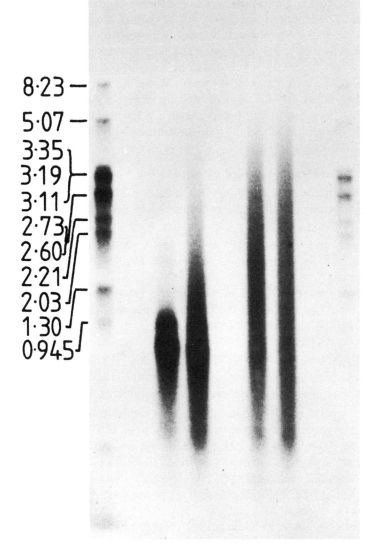

Figure 1. Aliquots of first and second strand reactions (5000 c.p.m.) were ethanol-precipitated after the addition of 2 μg carrier DNA, dried and resuspended in 20 μl alkaline gel buffer (30 mM NaOH, 1 mM EDTA). After addition of bromcresol green-glycerol loading dye, the samples were electrophoresed through a 1.2% alkaline agarose gel (cast in the same buffer) at 150 mA for 3 h with buffer recirculation, blotted to Whatman DE81, dried and autoradiographed. **A** refers to *Aplysia* atrial gland cDNA, while **L** refers to monkey hypothalamus cDNA; **1** and **2** refer to first and second strand reactions. As noted in the text, the label in the second strand reaction reflects second strand cDNA synthesis. **M** refers to marker lanes containing 5000 c.p.m. of *Hind*III-digested Ad2 DNA which was end-labeled by a kinase reaction. The two specific bands in lane A1 (clearly visible above the smear of heterogeneous reaction products) represent transcripts of mRNAs encoding two very prevalent neuropeptide precursors called A and B (16). The quality of second strand transcripts produced is particularly clear when lanes **A1** and **A2** are compared.

from the analogous reaction described in Chapter 2, since it incorporates RNase H to partially degrade RNA in the DNA-RNA duplex, and so provide oligoribonucleotides as primers for second strand synthesis. Additionally, there is no S1 nuclease step. The length of the cDNA synthesised in the first strand and second strand reactions can be determined by electrophoresis on an alkaline agarose gel as described in Section 2.5.3 of Chapter 2. The DNA can be transferred from the gel onto a sheet of DE81, by blotting as one would for a rapid 'Southern Transfer'. Lay the gel on a sheet of 'Saran Wrap' or 'Cling Film', and cover it with a sheet of DE81 paper, a wad of blotting paper, a glass plate and a weight. Remove the DE81 paper after 2 h and dry on a gel dryer before performing autoradiography. Neutralisation of the gel is not necessary.

2.3 Preparing the Double-stranded cDNA for the Vector

The subsequent steps are carried out in a similar manner to the analogous reactions in the previous chapter. The procedures are detailed in *Table 3*. They involve:

(i)   Filling in any 'ragged ends' on the cDNA so that chemically synthesised linkers can be ligated to it, and create *Eco*RI sites on the end of the molecules.

(ii)   Protecting internal *Eco*RI sites within the DNA from subsequent cleavage with *Eco*RI by modifying the DNA with the *Eco*RI methylase.

(iii)   Ligating linkers to the cDNA.

(iv)   Cleaving with *Eco*RI to remove all but the terminal unit of the oligomeric linker addition and generate cohesive *Eco*RI termini on the molecules.

The linkers are radiolabelled before being added to the cDNA. This is done by carrying out a 50 μl reaction catalysed by T4 polynucleotide kinase as follows:

(i)   Set up a 50 μl reaction containing 5 μl of dephosphorylated *Eco*RI linkers (5 μg), 5 μl of freshly prepared 10 x kinase buffer (100 mM Tris-HCl pH 7.5, 100 mM MgCl$_2$, 100 mM DTT, 10 mM ATP), 36 μl of water, 1 μl of high specific activity gamma-[^{32}P]ATP (10 μCi) and 3 μl T4 polynucleotide kinase (15 units, Boehringer) where 1 unit is the enzyme activity catalysing the incorporation of 1 nmol of ^{32}P in acid-precipitable products within 30 min at 37°C.

(ii)   Incubate for 1 h at 37°C and then freeze 5 μl aliquots at -20°C.

The success of the kinase reaction should be assessed by ligating an aliquot of the linkers, and subsequently digesting a portion of this ligation reaction with *Eco*RI. The reaction products can be examined by electrophoresis on a 10% polyacrylamide gel as suggested in Section 2.5.6 of Chapter 2 or by electrophoresis on a 6% acrylamide-7 M urea sequencing gel, in which case the bromphenol blue marker dye should be electrophoresed halfway down the gel. The ligation reaction should produce a 'ladder' of linker oligomers. *Eco*RI digestion of the ligation should reduce all linkers to monomer length. Remember to heat-inactivate the ligase (5 min, 70°C) before *Eco*RI digestion.

Table 3. Preparing the cDNA for the Vector.

A. *Filling in reaction*

1. Resuspend the cDNA in 25 μl from the last step in *Table 2*. Add 1.0 μl of 5 x d(ACGT)TPs[a], 1.0 μl 2 mg/ml BSA (Pentex, nuclease free), 9 units of T4 DNA polymerase[b] and 3.0 μl of 10 x T4 polymerase buffer[c] to give a final reaction volume of 30 μl.

2. Incubate at 37°C for 30 min and then add 30 μl of phenol and 30 μl chloroform. Shake and then centrifuge to separate the phases. Re-extract the phenol phase with 30 μl of TE.

3. Ether extract and ethanol precipitate as in step 3 of *Table 1*. Resuspend the final pellet in 4.8 μl of water.

B. *Methylation of EcoRI sites*

1. To the cDNA resuspend in 4.8 μl H_2O add 1.0 μl of SAM (luCi)[d], 2.0 μl of 5 x methylase buffer[e], 2.0 μl of 2 mg/ml BSA; and 0.2 μl of *Eco*RI methylase (26 units[f]/μl).

2. Incubate at 37°C for 20 min. Extract the reaction mixture with phenol and chloroform, re-extract the organic phase, pool and ether extract the aqueous phases, and then precipitate the nucleic acids as in steps 4 – 8 of *Table 1*.

3. Pellet the nucleic acids by centrifugation. Pour off the supernatant and dry the pellet under vacuum.

C. *Ligation of linkers to cDNA*

1. Resuspend the pellet in 5 μl of 1 x kinase buffer[g], 5 μl of kinased linkers (500 ng)[h], and 0.2 μl of 22 mM ATP.

2. Add 0.5 μl T4 DNA ligase (400 units[i]/μl) and incubate at 14°C overnight.

D. *Digestion of linkered cDNA to remove excess linkers*

1. Heat the ligated cDNA/linkers at 65°C for 10 min. Centrifuge briefly in a microcentrifuge to collect the condensate. Add 1.5 μl of 1 M Tris-HCl pH 7.5, 1.0 μl of 1 M NaCl (filtered through a nitrocellulose membrane) and 5.6 μl of water.

2. Add 140 units[j] of *Eco*RI and incubate for 2 h at 37°C.

3. Add 30 μl of TE and extract with phenol/chloroform. Re-extract the organic phase twice with 50 μl of TE. Extract the aqueous phase three times with ether and then add 15 μl of 2 M sodium acetate and 450 μl of ethanol. Chill to − 70°C and centrifuge in a microcentrifuge. Pour off the supernatant and dry the pellet under vacuum.

4. Resuspend the pellet in 20 μl of 0.4 M NaCl, 10 mM Tris-HCl pH 8.0, 1 mM EDTA (STE) and 2 μl bromophenol blue/glycerol. Fractionate the cDNA from the excess linkers by gel filtration on a 1 ml Ultrogel AcA34 (LKB) column in a drawn out, siliconised Pasteur pipette plugged with siliconised glass wool. The Ultrogel is equilibrated and the column is run in STE. Collect 100 μl fractions. Determine the radioactivity of the fractions and pool the peak of cDNA[k].

---

[a]5 x d(ACGT)TP is described in footnote [c] of *Table 1*.

[b]1 unit of T4 DNA polymerase catalyses the incorporation of 10 nmol of total nucleotide into an acid-insoluble product in 30 min at 37°C. The authors use enzyme supplied by PL Biochemicals.

[c]10 x T4 polymerase buffer is 330 mM Tris-acetate pH 7.9, 660 mM potassium acetate, 100 mM magnesium acetate, 1 mg/ml BSA (Pentex, nuclease free), 5 mM DTT.

[d]SAM is S-adenosyl-L-[methyl-^{3}H]methionine from Amersham, product number TRK 236; specific activity 15 Ci/mmol. Other commercial sources of SAM are often contaminated with S-adenosyl homocysteine which inhibits the methylase.

[e]5 x methylase buffer is 500 mM Tris-HCl pH 8.0, 5 mM EDTA pH 8.0.

[f]1 unit of *Eco*RI methylase is the amount of enzyme required to protect 1 μg of lambda DNA in 1 h at 37°C in a 10 μl reaction mixture against cleavage by *Eco*RI restriction endonuclease. The authors use enzyme supplied by New England Biolabs.

[g]Kinase buffer is described in the text in Section 2.3.

[h]Kinasing linkers is described in the text in Section 2.3.

[i]1 unit of T4 DNA ligase is defined as the amount of enzyme required to give 50% ligation of *Hind*III digested lambda DNA in 30 min at 16°C in 20 μl at a 5′ termini concentration of 0.12 mM (~330 μg/ml). The authors use enzymes purchased form New England Biolabs.

[j]One unit of *Eco*RI is the amount of enzyme required to digest 1 μg of lambda DNA in 1 h at 37°C in the assay buffer. The authors use high concentration enzyme purchased from Boehringer.

[k]A detailed protocol for the gel filtration is also given in Section 2.5.8 of the previous chapter.

Table 4. Ligation of Vector DNA to cDNA.

A. *Digestion of λ Insertion Vector DNA*

1. Digest about 50 μg of vector DNA in a 300 μl reaction with a 6 μl *Eco*RI (140 units/ml) for 2−3 h at 37°C.
2. Extract, with phenol and chloroform, back-extract, ether extract and ethanol precipitate at room temperature, but otherwise as described in steps 4−8 of *Table 1*.
3. Pellet the DNA by centrifugation, pour off the supernatant, allow the pellet to dry and then resuspend it in 50 μl of TE.

B. *Ligation of Vector to cDNA*

1. Aim for a 2:1 molar ratio of vector to cDNA, i.e., approximately 5 μg of phage arms and 100 ng of cDNA (assuming a mean size of 2 kb for the cDNA). Larger ratios are acceptable. Ideally, the DNA concentration in the ligation should be at least 1 mg/ml.

    Co-precipitate the cDNA and vector arms in ethanol. Chill the mixture on a dry ice-isopropanol bath. Pellet the DNA by centrifugation. Pour off the supernatant and allow the pellet to dry before resuspending it in 3.5 μl of water. Add 0.5 μl of 10 x ligase buffer[b], 0.5 μl of 10 mM ATP and 0.5 μl (200 units) T4 DNA ligase. It may be easier to do a double-size ligation (i.e., 10 μg arms and 200 ng cDNA) as it is easier to resolubilise the DNA in a volume of 7 μl. Incubate the reaction mixture overnight at 14°C.
2. Dilute the reaction mixture to 50 μl with TE and heat for 5 min at 65°C. The DNA is now ready to be packaged.

[a]10 x *Eco*RI reaction buffer is 1 M Tris-HCl pH 7.2, 500 mM NaCl, 50 mM MgCl$_2$, 20 mM 2-mercaptoethanol.
[b]10 x ligase buffer is 600 mM Tris-HCl pH 7.5, 100 mM MgCl$_2$, 100 mM 2-mercaptoethanol, 10 mM EDTA, 1 mM ATP.

2.4 Ligation of Vector DNA to cDNA

The digestion of λ insertion vector DNA with *Eco*RI and its subsequent ligation to the cDNA is described in *Table 4*.

3. IN VITRO PACKAGING AND PLATING OUT OF HYBRID PHAGE

The recombinant DNA from the final step in *Table 4* is now ready for packaging into phage particles *in vitro*. This chapter will not consider these aspects, since *in vitro* packaging is described in reference 13 and the plating out of recombinant phages carrying cDNA is described in the previous chapter from Section 2.5.12 onwards.

4. CONCLUDING REMARKS

A number of factors should be considered when choosing a host-vector system for cDNA cloning. Recombination systems of both host and vector origin may produce sequence rearrangements during initial plating or amplification. *hfl* is *recA*$^+$ and λgt10 and λgt11 are *red*$^+$. For instance, sequence instability observed when mRNAs composed of short tandemly repeated sequences like collagen and *Plasmodium falciparum* surface antigens are cloned in a *red*$^+$ vector like λgt11, can be alleviated by use of a *red*$^-$ derivative (14; J.Scaife, personal communication). These are, however, extreme cases. It may be preferable to use a *red*$^-$ vector like NM641 (8) or NM1150 (4) and even a *recA*$^-$ host like NM553 (which is *hfl⁻* r⁻ m⁺; N.Murray, personal communication). Apparent sequence rear-rangements due to reverse transcriptase artefacts have also been observed (15).

As discussed in detail in Chapter 2, λgt10 was constructed to provide a vector optimised for cloning typically sized eukaryotic mRNA molecules. If cloning of a much larger mRNA is to be attempted (>5 kb), use of an alternative vector capable of accepting larger inserts, like NM1149 or its *red⁻* derivative, NM1150, is preferable.

Often, the first clone isolated from a library screen is a partial cDNA clone (when insert size is compared to that of the corresponding mRNA on a Northern blot). Longer clones are normally isolated by rescreening the library using the short clone as a probe. Unfortunately, this can select for clones containing an ad-ditional (unrelated) insert. This can occur during linker addition (the problem is minimised by the large molar excess of linkers) or during ligation of vector to cDNA (where double inserts in a small fraction of clones are probably unavoidable). Certainly a cDNA clone found to hybridise to multiple mRNAs on Northern blots (particularly if it contains internal *Eco*RI sites) should be viewed with suspicion. Wherever possible, it should be shown that independently isolated clones have identical restriction maps. Sequences at the 5' end of mRNAs can be determined or verified by primer extension (if the RNA is sufficiently abundant) or comparison to genomic clones.

5. ACKNOWLEDGEMENTS

We wish to thank Hiroto Okayama and Paul Berg (Stanford) for providing the second strand reaction conditions prior to publication. This protocol evolved through the efforts of a number of individuals in Richard Axel's laboratory (Michael Palazolo, Scott Zeitlin, Toni Claudio and Greg Lemke) where JFJ was supported by a National Institutes of Health postdoctoral fellowship. We would also like to thank Barbara Wold (Caltech), Rick Young (Stanford) and Noreen Murray (Edinburgh) for advice and strains. CJW holds a research studentship from the Science and Engineering Research Council. The work of CJW and JFJ is supported by the Cancer Research Campaign and the Medical Research Coun-cil, respectively.

6. REFERENCES

1. Okayama,H. and Berg,P. (1982) *Mol. Cell Biol.*, **2**, 161.
2. Efstratiadis,A., Kafatos,F.C., Maxam,A.M. and Maniatis,T. (1976) *Cell*, **7**, 279.
3. Gubler,U. and Hoffman,B.J. (1983) *Gene*, **25**, 263.
4. Murray,N.E. (1983) in *Lambda II*, Cold Spring Harbor Laboratory Press, New York.
5. Williams,B.G. and Blattner,F.R. (1980) in *Genetic Engineering: Principles and Methods*, Set-low,J.K. and Hollaender,A. (eds.), New York, Plenum.
6. Scherer,G., Telford,J., Baldari,C. and Pirrotta,V. (1981) *Dev. Biol.*, **86**, 438.
7. Lathe,R. and Lecocq,J.-P. (1977) *Virology*, **83**, 204.
8. Murray,N.E., Brammar,W.J. and Murray,K. (1977) *Mol. Gen. Genet.*, **150**, 53.
9. Stanley,K.R. and Luzio,J.P. (1984) *EMBO J.*, **3**, 1429.
10. Ruther,U. and Muller-Hill,B. (1983) *EMBO J.*, **2**, 1791.
11. Chirgwin,J.M., Przybyla,A.E., Macdonald,R.J. and Rutter,W.J. (1979) *Biochemistry*, **18**, 5294.
12. Cathala,G., Savouret,J.-F., Mendez,B., West,B.L., Karin,M., Martial,J.A. and Baxter,J.D. (1983) *DNA*, **2**, 329.

13. Maniatis,T., Fritsch,E.F. and Sambrook,J. (1982) *Molecular Cloning: A Laboratory Manual*, Cold Spring Harbor Laboratory Press, Cold Spring Harbor, NY.
14. Hall,R., Hyde,J.E., Goman,M., Simmons,D.L., Hope,I.A., Mackay,M., Scaife,J., Merkli,B., Richle,R. and Stocker,J. (1984) *Nature,* **311**, 379.
15. O'Hare,K., Breathnach,R., Benoist,C. and Chambon,P. (1979) *Nucleic Acids Res.,* **7**, 321.
16. Scheller,R.H., Jackson,J.F., McAllister,L.B., Rothman,B.S., Mayeri,E. and Axel,R. (1982) *Cell,* **32**, 7-22.

CHAPTER 4

Immunological Detection of Chimeric β-Galactosidases Expressed by Plasmid Vectors

MICHAEL KOENEN, HANS-WERNER GRIESSER and BENNO
MÜLLER-HILL

1. INTRODUCTION

In recent years various expression vectors have been constructed and used to clone procaryotic genomic or eucaryotic cDNA. These vectors allow the identification of cloned DNA using methods for the specific detection of the expressed protein. In some cases the expression vector contains just a promoter for transcription, a Shine Dalgarno sequence and a start codon for translation. In other cases the vector contains a carrier gene, into which foreign DNA can be cloned.

Since our first description of an active β-galactosidase chimera (1), many cloning systems have been described which utilise the fact that the 25 N-terminal residues of β-galactosidase can be replaced by peptides of any size and origin without destroying its activity (2 − 11). In this chapter we shall describe cloning vectors of this type and the procedures for working with them that we have developed in our laboratory. Our vectors permit cloning of foreign DNA in either the 5′ or 3′ end of the *lac Z* gene (5,12). DNA cloned into these sites may be expressed as part of a chimeric β-galactosidase. Cloning of DNA into the 5′ end of the *lac Z* gene may lead to fusions where the insert codes for the N-terminal part of active β-galactosidase. Cloning in the 3′ end of the *lac Z* gene may lead to fusions where the C terminus of active β-galactosidase is coded by the inserted DNA. DNA of known sequence may be inserted directly as a restriction fragment into the proper reading-frame of the *lac Z* gene. On the other hand, if the sequence is unknown, the DNA may be inserted using dG/dC tailing, in which case few recombinants will have the proper translational reading frame and orientation. We find that under appropriate conditions most, though not all, chimeric β-galactosidases are 90% stable. We will describe how the use of β-galactosidase as a carrier enzyme offers the possibility of a general purification procedure for all chimeric antigens. It also allows the detection of positive clones by their β-galactosidase activity (5,6).

2. CLONING INTO THE 5′ END OF THE LAC Z GENE

Plasmid pUK270 (*Figure 1*) carries the entire *lac* operon: the *lac Z* gene with

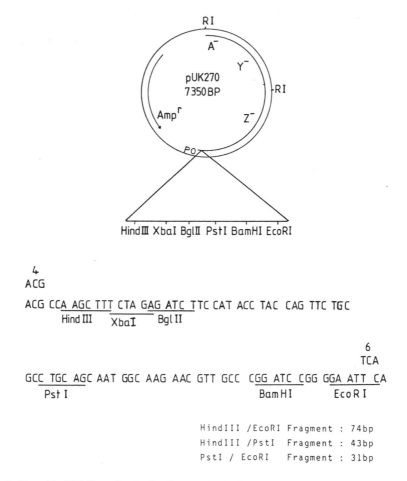

Figure 1. Plasmid pUK270 carries the *lac Z* gene, the *lac Y* gene and parts of the *lac A* gene. A frame shift mutation in the polylinker region shifts the plasmid to *lac Z⁻*.

a frame shift mutation in a polylinker in the promoter proximal part; the *lac Y* gene and part of the *lac A* gene. Bacteria containing pUK270 cannot grow on lactose minimal agar plates if their chromosomal *lac* operon is deleted. The host we use, F11 *rec A*, carries a (*lac pro*) deletion and a F'*lac pro* episome. The episome expresses lac repressor from an I^{q1} promoter and carries a deletion of the entire *lac Z* gene. It is active for *lac* permease. We emphasise here that this host cannot be replaced by just any host which can be well transformed. The host has to overproduce *lac* repressor to keep the plasmid's *lac Z* genes repressed when not induced. The presence of large amounts of *lac* repressor does not allow full expression of the plasmid *lac Z* genes even in the presence of lactose or the inducer isopropyl-1-thio-β-D-galactosidase (IPTG). This is necessary to preserve the stability of the *lac* operon of the plasmid. Overexpression of β-galactosidase is a growth disadvantage and leads to the accumulation of plasmid mutants which have defects in the insert and the *lac* operon. F11 *rec A* bacteria

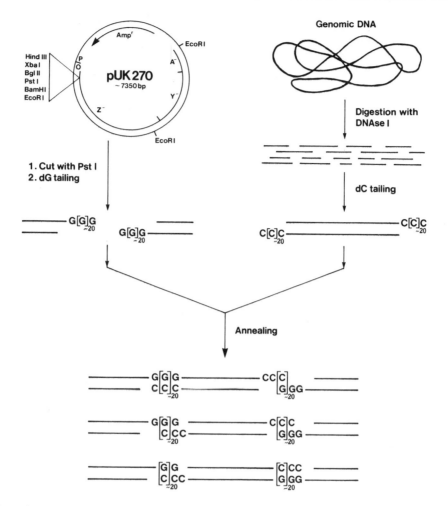

Figure 2. Cloning of genomic DNA in the the *Pst*I site of pUK270 after dG/dC tailing. Annealing either at the 5′ end or at the 3′ end of the insert allows annealing of the tails in every possible reading frame.

containing the plasmid pUK270 show a pale blue colour on rich media plates containing the inducer IPTG and the histochemical stain 5-bromo-4-chloro-3-indolyl-β-D-galactoside (X-Gal) indicating that some β-galactosidase is expressed. We observe the same phenomenon with the similar vectors pUK230 and pUK250 constructed previously (5,13). A weak restart or a frame shift suppressor may be the reason for this residual activity. Whatever the reason, this activity does not harm the use of the vectors. Plasmid pUK270 permits cloning of DNA fragments of known sequence into one of the six unique restriction sites of the polylinker, but the major use is cloning of *random* DNA fragments into the *Pst*I site after dG/dC tailing.

Recombinant clones can be selected as *lac* positives on lactose minimal or on rich media plates containing IPTG and X-Gal. The efficiency of transformation, however, is diminished by an order of magnitude if transformants are plated directly on minimal lactose agar. For this reason we first plate transformants on rich media plates containing ampicillin. We select *lac* positives after colony transfer onto lactose minimal plates. It is also possible to screen recombinants as dark blue colonies on rich media plates containing IPTG, X-Gal and ampicillin. The number of *lac*+ colonies varies between 5% and 10% of the total transformants when random DNA fragments are cloned after dG/dC tailing. The exogenous DNA fragments anneal in both orientations in the six possible reading frames with the vector DNA (*Figure 2*).

Half of the clones are in the wrong orientation. Of those clones which carry the insert in the proper orientation 1/3 will be in phase at the 5' end and again only 1/3 of these will maintain the reading frame at the 3' end of the insert. In our previous experiments we observed that the frequency of lactose positives is often higher than the expected 5% (5,6). This increase of *lac*+ clones can be explained in two ways. The first possibility would be that the inserted DNA carries its own initiator codon for translation. This is supported by the high frequency with which we were able to isolate the 5' end of the *lac I* gene. The second possibility is that small DNA fragments may carry more than one open reading frame.

The polylinker of pUK270 permits direct chemical sequencing from both sides of inserted DNA cloned in either the *Pst*I site or the *Bgl*II site (14,15). Alternatively, the *Hind*III and the *Eco*RI site of the polylinker allow easy recloning into the M13 mp8 or mp9 system (16) for subsequent use of enzymatic DNA sequence analysis (17).

3. CLONING IN THE 3' END OF THE LAC Z GENE

We have also constructed a series of vectors which allow the cloning of DNA into the 3' end of the *lac Z* gene in such a way that active chimeric β-galactosidase is produced (12). These vectors may be used to clone cDNA if an active chimeric β-galactosidase is the desired product. The stop codon at the 3' end of cDNA makes it mandatory to use these vectors if one wants to have the β-galactosidase chimeras expressed. An inserted segment of cDNA from a particular clone may be recloned *via* its *Pst*I or other restriction sites. We have constructed vectors for all three translational reading frames (12). However, the vectors may also be used to establsh new cDNA banks. The size of such banks is only restricted by the frequency with which plasmid DNA can be introduced into *Escherichia coli*.

It is imperative to use a host that produces sufficient amounts of *lac* repressor to repress β-galactosidase production efficiently. We use the same host F11*rec A* as for the vector pUK270 discussed in Section 2. Both cloning systems could be adapted for use with phage lambda. Plasmid pUR290 (*Figure 3*), one of the vectors with differing translational reading frames in the cloning regions, carries the *lac Z* gene with its regulatory region and a polylinker inserted just before its stop codons (12). In contrast to pUK270, this expression vector has a *lac Z*+ genotype. Colonies of bacteria that harbour this plasmid and yet have the chromosomal *lac* operon deleted, have a dark blue colour on rich media plates containing IPTG and X-Gal. There are some other minor differences between

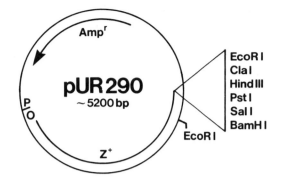

Figure 3. Plasmid pUR290 which carries the *lac* Z gene with a polylinker region in its 3' end. The plasmid is *lac* Z^+.

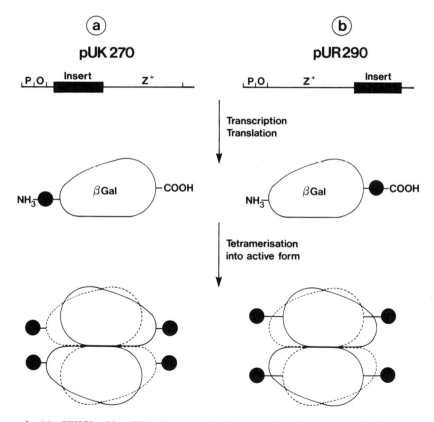

Figure 4. (a) pUK270 with a DNA fragment cloned in the polylinker region in the 5' end of the *lac* Z gene. The insert codes for the N terminus of a chimeric β-galactosidase. The chimera is a tetramer when active. (b) pUR290 with a DNA fragment cloned at the 3' end of the *lac* Z gene. The insert codes for the C terminus of the chimeric protein.

pUR290 and pUK270. pUR290 misses the *lac Y* gene and thus fails to allow selection of *lac* positive recombinants on lactose agar in a *lac X Y* deletion background. As we discussed in Section 2, the yield of *lac* recombinants when random DNA fragments are inserted by dG/dC tailing into pUK270 is 5−10%. A similar experiment with pUR290 leads maximally to 17% recombinants as only the 5' end of the exogenous DNA has to be inserted in the proper orientation and reading frame. A selection for enrichment of recombinant clones is not possible as both the recombinants and pUR290 are lactose positive. The insertion of DNA into the polylinker region of pUR290 enables one to perform direct chemical sequencing of the insert from both sides (14,15). The *Hind*III and *Bam*HI sites permit recloning of the insert into M13 mp8, mp9 for enzymatic DNA sequencing (16,17).

4. EXPERIMENTAL APPROACHES

4.1 Materials

4.1.2 *Sheets for Absorbing Antibodies*

We have used polyvinyl chloride (PVC) sheets for this purpose. These were supplied by Fa. Benneke Hannover-Vinnhorst and the trade name of the material is Bennecor, Acetella glasklar. The sheets should be covered on one side with a thin sheet of paper to protect the surface of the polymer. They should be highly flexible in order to facilitate placing them onto lysed bacterial colonies on an agar plate. The use of rigid material causes pressure to be exerted on the lysed colonies. This results in smearing of bacteria all over the sheet. This can be seen as a blue smear over the sheet. Avoid touching the surface of the polymer used for antibody binding. The sheets that we have used sometimes showed small grey structures that give rise to residual debris in the washing procedure. Protein binds unspecifically to such structures yielding a false positive reaction in the identifying step. Circular sheets, 8 cm in diameter, can be cut from the larger sheets with the help of a pair of compasses. In order to do this it is necessary to replace the pencil tip by the cutting edge of a scalpel or a razor blade. This instrument permits clean cutting. The use of other instruments often yields fine grains at the cutting edge of the polymer. This should be avoided since it leads to unspecific reactions in the final colour test.

4.1.2 *Reagents*

The buffer for antibody binding is 0.2 M $NaHCO_3$ pH 9.2. The washing solution consists of 150 mM NaCl, 10 mM $K-PO_4$ pH 7.2, 0.5% Triton X-100 (v/v), 0.5% bovine serum albumin (BSA). 2x TMSII buffer is 20 mM Tris HCl pH 7.5; 20 mM Mg acetate, 400 mM KCl, 0.2 mM EDTA, 100 mM β-mercaptoethanol. We have obtained X-Gal and IPTG from Bachem Inc. Bacterial plates were prepared according to (18) and IgG according to (19).

4.2 Coating of the Plastic Sheets with Antibodies

The coating of the PVC sheets with antibodies is carried out in clean sterile glass Petri dishes. In most experiments DEAE-purified IgG was used. We did not purify the IgG further, for example by affinity chromatography to antigen. 5 ml of binding buffer containing 90 µg IgG/ml is sufficient to coat 10 sheets. The procedure is carried out as follows:

(i)   Mark the sheets with a bench marker pen and place them on the surface of the antibody solution using a pair of forceps.

(ii)   Leave the sheets in contact with the antibody solution for 2 min at room temperature whilst shaking the Petri dish gently.

(iii)   Take the sheet out of the solution with a pair of forceps. Allow the excess antibody solution to run off the sheet before placing it into the wash solution in a vessel large enough to hold several sheets.

(iv)   Wash the sheets for about 5 min at room temperature. Avoid contact between the sheets. If the sheets touch each other they may lose their capacity to bind antibodies to their surface.

The washing procedure saturates the sheets with an excess of BSA which will block all the unspecific protein binding sites.

4.3 Lysis of Bacterial Colonies

Strictly speaking the word lysis is misleading, since the colonies are treated with chloroform which destroys the membranes but not the cell walls of *E. coli*. In fact, none of the procedures which cause lysis of the cell wall (for example, lysozyme treatment) can be used since they result in artifacts. It is therefore necessary to resort to using chloroform even though this is unsatisfactory in that it renders only a small amount of β-galactosidase available for the antibody reaction. There are two ways to prepare the colonies for treatment with chloroform.

(i)   Grow the colonies on lactose agar in glass Petri dishes at 37°C. Cool the plates for 1 h in a refrigerator before pouring 5 ml chloroform over the colonies. Incubate the colonies with chloroform for 2 min at room temperature and then pour it off. Place the plates at 37°C for sufficient time to allow residual chloroform to evaporate. Store the plates at 4°C until they are used.

(ii)   Grow up to 1000 colonies in plastic Petri dishes containing lactose agar. Place a nitrocellulose filter (Schleicher and Schüll) on the colonies for 2 min. Mark the orientation of the filter using a needle for the filter and a marker pen on the plate. Remove the filter from the plate using forceps and place it onto a fresh plate containing rich media in the agar. Incubate the plate for 1 h at 37°C. This procedure can be repeated several times with several filters. The colonies on the master plate can also be regrown overnight at 37°C. This technique (20) allows the same colonies to be tested with different antisera. The colonies on the filter are lysed with chloroform. Fill a glass Petri dish with 20 ml chloroform and dip one filter after the other into the chloroform for 1 min. Take the filter out using forceps and place it on filter paper for 3 min. Place it again on an agar plate and incubate it at 37°C for 5 min to remove residual chloroform. Store the plates at 4°C until ready for use.

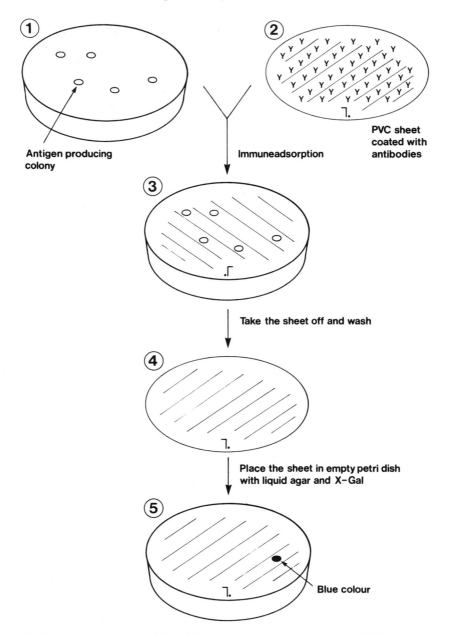

Figure 5. The immunoenzymatic procedure. Colonies are lysed with chloroform (**1**). PVC sheets are coated with antibodies (**2**) and placed on the lysed colonies for immunoadsorption (**3**). The sheets are cleaned with wash solution and placed in an empty Petri dish (**4**). Agar containing the histochemical stain X-Gal is poured over the sheets. Blue colour develops where chimeric β-galactosidase sticks to the sheets (**5**).

4.4 **Immunoadsorption**

Take the antibody coated and BSA saturated sheets out of the wash solution using forceps. If the bacterial colonies have been lysed on glass Petri dishes, it is necessary to drain all liquid from the sheets by letting their edges touch filter paper. Place the sheet onto the colonies using forceps and taking care to line up the mark on the sheet with a mark on the plate. Take care to prevent the formation of air bubbles between agar and plastic sheet! If bubbles have to be removed then try and do it carefully using gentle pressure from the tip of the forceps. Incubate for 1 h at 4°C to allow formation of immune complexes between the antigen and the antibodies.

If the colonies have been lysed on nitrocellulose filters, place the sheets directly onto the colonies on the filter without drying them. In this case it is necessary to use an excess of the wash solution in order to prevent the formation of air bubbles. Incubate the PVC sheet in contact with the nitrocellulose filter for 1 h at 4°C.

4.5 Wash Procedure

Remove the sheets from the plates and place them in a vessel containing wash solution. Most of the bacterial debris sticks to the sheets. To soak off the debris, the sheets should be covered completely by wash solution and incubated for 20 min at room temperature. All nonspecific adhering material is removed using a 20 ml syringe to spray the sheets with wash solution. For this purpose the sheets are placed with the dirty side up in a glass vessel with a large surface area. The vessel is held at an angle in order that the debris is washed away from the filter (*Figure 6*). Take care that the sheets do not slide down into the dirty wash solution. At the end of the procedure not a trace of a colony should be visible. If debris is still sticking on the sheets, soak them for another 5 – 10 min in fresh wash solution. Clean PVC sheets are soaked once more in fresh wash solution for 1 min and then placed in empty Petri dishes with the immune complexes on the upper side.

4.6 Identification of Antigen Producing Colonies

(i)   Melt agar (3 g/l of water) and store it at 40°C.

(ii)   Mix an equal volume of the liquid agar with 2x TMSII buffer pre-warmed to 40°C, and add X-Gal to a final concentration of 0.16 mg/ml.

(iii)   Cool to 37°C and pour about 15 ml on each sheet.

(iv)   When the agar has hardened, incubate the plate at 37°C in the dark. (X-Gal is light sensitive. The agar will turn yellowish if the lights are kept on!)

The X-Gal is hydrolysed by the β-galactosidase part of the antigen-β-galactosidase chimera bound to the antibody. The velocity of the colour reaction depends on the amount of chimeric β-galactosidase expressed, the quality of the antigen, and the amount and purity of the antibodies present. It may take between 2 and 20 h to see the colouration. Our experience using antibodies against one protein (*lac* repressor, chicken lysozyme) was that visible colour appeared after about 5 h (5,6). When the antibodies used were obtained against a mixture of proteins the colour became visible after 12 – 20 h. The position of the colonies that produce antigen recognized by the antibodies will be seen as a homogeneous blue coloured cloud comparable with the size of the lysed colony. Only positive clones

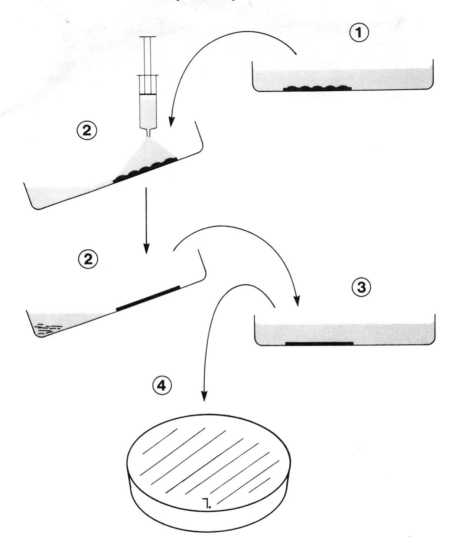

Figure 6. The wash procedure: soak the PVC sheets with the adhering colonies in the wash solution (**1**). Wash the material away using a syringe (**2**). Clean the sheet in fresh wash solution (**3**). Place the purified sheets into an empty Petri dish (**4**). See also Section 4.5.

show this homogeneous colour reaction. False signals or background will be seen if the washing procedure was insufficient. In such a case single small blue spots are visible at the border of the colony and the homogeneous cloud of blue colour typical for a positive result is not seen. It is imperative to verify all positive clones by fresh streaks from the colony of the master plate and by retransformation of its plasmid DNA.

4.7 Purification of the Chimeric β-Galactosidases

It is an advantage of the system that all antigens can be purified from extracts of *E. coli* cells to homogeneity in a single step. The purified antigen in the form of β-galactosidase-chimeras can then be used directly for immunisation. To purify chimeric β-galactosidases we have adapted the method of Steers *et al.* (21) which uses affinity chromatography on p-amino-β-D-thio-galactoside agarose. A variant of the method has been described by Ullmann (22).

(i)   Grow 200 ml of cells overnight in rich medium at room temperature to ensure stability of chimeric proteins.

(ii)   Resuspend 1 g cells in 1 ml 10 mM Tris-HCl pH 7.5, 10 mM Mg acetate, 200 mM NaCl, 0.1 mM EDTA, 1 mM DTE (buffer A).

(iii)   Sonicate the cell suspension three times for 1 min in an ice/NaCl bath (Branson Sonifier, 100 W power setting).

(iv)   Centrifuge at 40 000 *g* for 1 h to pellet the cell debris. The cell debris is assayed for β-galactosidase activity as described by Miller (18) and discarded if most of the activity is present in the supernatant.

(v)   Dilute the supernatant which contains 35 − 45 mg protein/ml in an equal volume of buffer A and apply it to a 1 ml p-amino-phenyl-β-D-thiogalactoside agarose (Sigma) column, equilibrated with buffer A.

(vi)   To remove nonspecifically bound protein, wash the column with 50 − 100 ml buffer A until there is no detectable protein in the column flow through.

(vii)   Elute the absorbed β-galactosidase chimera with three bed volumes of 100 mM sodium borate pH 10.05.

(viii) Dialyse immediately against 500 ml 40 nM Tris-HCl pH. 7.5, 1 mM $MgCl_2$, 0.1 mM EDTA, 1 mM DTE, 50% glycerol for 5 h and store at −20°C.

Chimeras purified according to this protocol have proved to be >90% homogeneous upon SDS-polyacrylamide electrophoresis (23). The samples were prepared for electrophoresis by incubation at 55°C for 30 min instead of boiling so as not to degrade the chimeric protein. The protein concentration may be determined by the method of Lowry relative to bovine serum albumin standards (24). β-Galactosidase activity was measured as described by Miller (18). All manipulations of the purification were carried out at 4°C.

5. GENERAL COMMENTS

Recombinant plasmids are stable when introduced into the host F11 *rec A* and produce chimeric β-galactosidases in large amounts. An exception to this rule are those chimeric β-galactosidases which are due to the re-initiation of protein synthesis in the DNA inserted at 5′ end of the *lac Z* gene. In most cases the length of the subunit of the chimera as seen on SDS gels is proportional to the length of the DNA insert. If the clones are selected with antibodies they have already been selected for a stable antigen. However, if we are interested to ensure the stability of the chimeras we grow the cells at room temperature. In general we have observed that inserts which are bordered by poly Gly and poly Pro are stable. Stability is generally good if the expressed peptide consists of a complete folding unit. It may be wise only to clone C-terminal fragments and not internal fragments at the 3′ end of the Z gene. Another problem that can arise is that of insolubility of the chimera when the foreign DNA expresses sequences which are highly lipophilic such as leader sequences or sequences that span membranes. It is also possible to use iodinated protein A to identify antigens expressed from DNA cloned in either the 5' or 3' end of the *lac Z* gene (25). We have used this method and the immunoenzymatic method described above in parallel and have obtained similar results.

The method is particularly attractive since one does not need to know the structure of the antigen. It is sufficient that antibodies directed against an antigen are available and the expression bank is of sufficient size. The cloning systems that we described could be adapted for use with lambda vectors and we see here possibilities for future developments and applications.

6. REFERENCES

1. Müller-Hill,B. and Kania,J. (1974) *Nature,* **249**, 561.
2. Roberts,T.M., Kacich,R. and Ptashne,M. (1979) *Proc. Natl. Acad. Sci. USA,* **76**, 760.
3. Guarente,L., Lauer,G., Roberts,T.M. and Ptashne,M. (1980) *Cell,* **20**, 543.
4. Casadaban,M.J., Martinez-Arias,A., Shapira,S.K. and Chou,J. (1983) in *Methods in Enzymology,* Vol. 100, Wu,R., Grossman,L. and Moldave,U. (eds), Academic Press, New York, p. 293.
5. Koenen,M., Rüther,U. and Müller-Hill,B. (1982) *EMBO J.,* **1**, 509.
6. Rüther,U., Koenen,M., Sippel,A.E. and Müller-Hill,B. (1982) *Proc. Natl. Acad. Sci. USA,* **79**, 6852.
7. Gray,M.R., Colot,H.V., Guarente,L. and Rosbash,M. (1982) *Proc. Natl. Acad. Sci. USA,* **79**, 6598.
8. Zabeau,M. and Stanley,K.K. (1982) *EMBO J.,* **1**, 1217.
9. Weinstock,G.M., Aprhys,C., Berman,M.L., Hampar,B., Jackson,D., Silhavy,T.J., Weisemann,J. and Zweig,M. (1983) *Proc. Natl. Acad. Sci. USA,* **80**, 4432.
10. Weis,H.J., Enquist,L.W., Salstrom,J.S. and Watson,R.J. (1983) *Nature,* **302**, 72.
11. Roy,A., Danchin,A., Josef,E. and Ullmann,A. (1983) *J. Mol. Biol.,* **165**, 197.
12. Rüther,U. and Müller-Hill,B. (1983) *EMBO J.,* **2**, 1791.
13. Rüther,U. (1983) Thesis, Cologne.
14. Maxam,A.M. and Gilbert,W. (1980) in *Methods in Enzymology,* Vol. 65, Grossman,L. and Moldave,K. (eds), Academic Press, New York, p. 499.
15. Rüther,U. (1982) *Nucleic Acids Res.,* **10**, 5765.
16. Messing,J. and Vieira,J. (1982) *Gene,* **19**, 269.
17. Sanger,F., Nicklen,S. and Coulson,A.R. (1977) *Proc. Natl. Acad. Sci. USA,* **74**, 5463.
18. Miller,J.M. (1972) *Experiments in Molecular Genetics,* Cold Spring Harbor Laboratory Press, Cold Spring Harbor, NY.
19. Porter,R.R. (1959) *Biochem. J.,* **73**, 119.
20. Hanahan,D. and Meselson,M. (1983) in *Methods in Enzymology,* Vol. 100, Wu,R., Grossman,L. and Moldave,K. (eds), Academic Press, New York, p. 333.
21. Steers,E., J., Cuatrecasas,P. and Pollard,H.B. (1971) *J. Biol. Chem.,* **246**, 196.
22. Ullmann,A. (1984) *Gene,* **29**, 27.
23. Laemmli,U.K. (1970) *Nature,* **227**, 680.
24. Lowry,O.H., Rosebrough,N.J., Farr,A.L. and Randall,R.J. (1951) *J. Biol. Chem.,* **193**, 265.
25. Kemp,D.J., Coppel,R.L., Cowmann,A.F., Saint,R.B., Brown,G.V. and Anders,R.F. (1983) *Proc. Natl. Acad. Sci. USA,* **80**, 3787.

CHAPTER 5

The pEMBL Family of Single-stranded Vectors

LUCIANA DENTE, MAURIZIO SOLLAZZO, COSIMA BALDARI,
GIANNI CESARENI and RICCARDO CORTESE

1. INTRODUCTION

Single-stranded (s.s.) vectors are used in many laboratories for DNA sequencing, site directed mutagenesis, S1 mapping and as s.s. probes for hybridisation. The most widely used vectors are the M13 derivatives developed by Messing and colleagues (1,2). In practice however there are certain limitations to the generality of the use of M13 derived vectors. It has, for example, been repeatedly observed that large inserts tend to be unstable. Furthermore in many cases it is necessary to clone a segment of DNA in a plasmid designed for a specific purpose. This could, for example, be a shuttle vector designed for transferring DNA between eukaryotic and prokaryotic cells and carrying genes allowing its selection in either system. On the other hand it could be designed as an expression vector in either system. It would be convenient, therefore, if any of these specialised plasmids could be converted into a single-stranded DNA form for DNA sequencing experiments, site-directed mutagenesis, etc. Dotto *et al.* have cloned a segment of the f1 genome containing all the cis-acting elements required for DNA replication and morphogenesis into the *Eco*RI site of pBR322 (3,4). These authors showed that if cells containing such plasmids were superinfected with f1 phage, the virion capsids secreted into the culture medium contain either f1 s.s. DNA or pBR322 s.s. DNA at about the same frequency. We have exploited this property in order to construct a variety of new cloning vectors.

2. THE pEMBL FAMILY

Vectors of pUC or Mp families of vectors (1,2) are very versatile because they allow one to insert foreign DNA sequences at a variety of restriction endonuclease sites within a short polylinker DNA contained in the coding sequence of the α-peptide of the enzyme β-galactosidase. This offers the possibility of inserting DNA fragments in a selected orientation or in a specific translational reading frame in the nucleotide sequence encoding the α-polypeptide. Bacterial colonies containing recombinant clones can be directly identified by simple colour tests on indicator plates without the need of replica plating. In order to expand the versatility of these plasmids we have cloned the 1300 bp *Eco*RI-*Eco*RI segment of the f1 genome from the plasmid pD4 [described by Dotto *et al.* (3)] into the unique *Nar*I site present in all pUC plasmids. The resulting plasmids [the pEMBL family

(5)] maintain all the pUC specific functions. They have the same origin of replication. They each determine resistance to the antibiotic ampicillin. Furthermore they synthesise the functional α-peptide of β-galactosidase and so produce blue colonies when plated on the appropriate indicator stain in the presence of the chromogenic substrate 5-chromo-4-chloro-3-indolyl-β-D-galactosidase (X-Gal). In general the gene for the α-peptide is inactivated in recombinant plasmids, resulting in the formation of white plaques on plates containing X-Gal. In addition, however, the pEMBL plasmids contain the intact origin of replication of f1 that can be activated when cells are superinfected with f1 phage. This results in the production of s.s. plasmid DNA that can be packaged and secreted into the culture medium as a virion-like particle. The orientation of the 1300 bp segment of f1 DNA determines which of the two strands is encapsidated in the viral particle and accordingly the vectors have been named pEMBL(+) and pEMBL(−). In the pEMBL(+) vectors, the antisense strand of the β-galactosidase gene is present in the virions produced by superinfected cells. In the pEMBL(−) vectors the coding strand of the β-galactosidase gene is found in the virion. Physical maps of the pEMBL plasmids are shown in *Figure 1*.

3. ALTERNATIVE SINGLE-STRANDED PLASMID VECTORS FOR SPECIAL APPLICATIONS

The possibility of preparing s.s. DNA from any plasmid is an advantage that would also be useful in other plasmids with additional specialised features. In principle any existing plasmid could be made into a 'pEMBL type' plasmid by insertion of the restriction fragment carrying the replication origin of f1. We have explored this possibility by constructing specialised yeast shuttle vectors and *E. coli* expression vectors of this type.

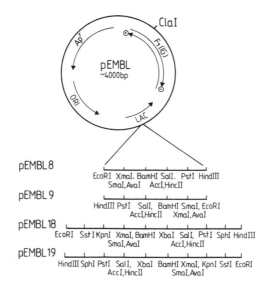

Figure 1. Physical map of pEMBL plasmids.

The pEMBLY vectors are a family of plasmids derived from pEMBL9 by inserting a number of yeast DNA fragments into its unique *Cla*I site. The yeast DNA carries genetic determinants which can be used as selective markers in appropriate mutant strains of yeast (Baldari and Cesareni, in preparation). In addition some of the yeast DNA fragments can support plasmid replication in *S. cerevisiae per se*. Yeast vectors without DNA replication origins transform yeast at low frequency (6). These are the so-called integration vectors, which yield stable transformants by undergoing recombination with homologous sequences within the chromosome. Yeast plasmids with yeast replication origins give transformation efficiencies that are up to three orders of magnitude greater than the integration vectors. The replication origin can be supplied either from the endogenous yeast 2 μ plasmid or as *ars* sequences (7). The latter sequences occur in the DNA of many eukaryotes and appear to act as replication origins in yeast (8). The pEMBLY family includes plasmids of all these types and vectors are available that permit the selection of yeast transformants by complementation of the *LEU*2, *URA*3 or *TRP*1 genes. *Figure 2* shows the physical maps of the pEMBLY plasmids. The unique restriction endonuclease sites located in the sequence encoding the α-peptide of β-galactosidase can be used as cloning sites.

	cloning sites	Other unique sites
pEMBL Yi 22 / pEMBL Yi 21 (opposite orientation) — ura3; Av Sm St Ap N E V P	BamHI, EcoRI, Hind III, Sal I	Apal, Nco I, EcoRV
pEMBL Ye 23 / pEMBL Ye 24 (opposite orientation) — ura 3 / 2μ; Av Sm St Ap N/P ES t P Hp Av X, EV	BamHI, Hind III, Sal I	Apal, Nco I, EcoRV
pEMBL Yr 25 — trp 1; E X M H St Bg E, EV P	Ava I, BamHI, Sal I, Sma I	EcoRV, Bgl II, Mst I, Stu I, Xba I
pEMBL Yi 27 — leu 2; Av Hp Bst CK E EV, Na	BamHI, Hind III, Pst I, Sal I, Sma I	Bst EII, Cla I, EcoRV, Kpn I, Nar I
pEMBL Ye 30 — 2μ / leu 2; X Av Hp P EV E KC Bst Hp Av Na	BamHI, Hind III, Sma I	Bst EII, Cla I, EcoRV, Kpn I, Nar I
pEMBL 9 — F1 ori, lacZ, ori, bla; Sm C E B Sa Av H		

Figure 2. Diagrams representing the physical and genetic map of pEMBLY plasmids. The position of the restriction endonuclease recognition sites was deduced from the DNA sequence of pEMBL9 (5), *URA*3 (11), *TRP*1 (12) and 2μ (13). We have assumed, without any direct sequence analysis, that no rearrangement or deletion occurred at the boundaries between the DNA fragments after the various enzymatic treatments. Unique sites in the α-peptide coding sequence have been checked by restriction endonuclease digestion of the appropriate plasmid DNA and analysis of the product by electrophoresis on 1% agarose gel. Restriction enzyme abbreviations are as follows: E = *Eco*RI, N = *Nco*I, EV = *Eco*RV, St = *Stu*I, Hp = *Hpa*I, M = *Mst*I, Bg = *Bgl*II, X = *Xho*I, K = *Kpn*I, Bst = *Bst*II and N = *Nar*I. Arrows indicate the direction of transcription or replication.

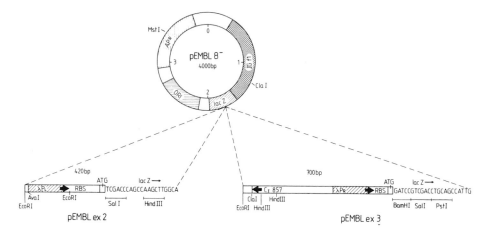

Figure 3. Physical maps of expression vectors pEMBL ex1 and pEMBL ex2. The nucleotide sequence of the polylinker in the two vectors was deduced from the DNA sequence of the fragments that have been assembled taking into account the enzymatic steps used in the construction. RBS indicates the ribosome binding sites of the MS2 replicase gene and of the λ*cro* gene.

The second type of specialised single-stranded 'pEMBL type' plasmid vector was designed as an expression system. We wished to make an expression vector that would direct the synthesis of a protein of interest from an initiator AUG supplied by the vector. Subsequently one would be able to make precise deletions of unwanted sequences by the procedure of annealing a specific oligonucleotide to the single-stranded vector, repairing this *in vitro* and then letting the mutant and wild-type strands segregate between progeny virions. This approach to *in vitro* mutagenesis is described in Chapter 8 of this book. The method avoids random enzymatic deletions and lengthy sequence analysis of many different clones. Once the proper construction has been obtained that overexpresses the protein, oligonucleotide directed mutagenesis could be conveniently performed on this construct. The plasmid carrying the mutant coding sequence could then be used immediately to produce large amounts of the corresponding altered gene product in *E. coli*.

The plasmids that we have constructed (Sollazzo and Cesareni, in preparation) are derived from two plasmids containing either the lamba P_L or P_R promoter which direct transcription in the direction of a sequence containing the ribosome binding site and the initiator AUG of the replicase gene of MS2 phage (9) and of the lambda *cro* gene (10) respectively. These fragments have been inserted in the polylinker region of pEMBL8− in such a way that a universal primer complementary to the ribosome binding site region can be used to sequence across the AUG (*Figure 3*). The initiator AUG was inserted in frame with the sequence encoding the α-peptide of β-galactosidase so as to maintain the blue-white screening for insertion on X-Gal indicator plates. Convenient unique restriction endonuclease sites are located immediately downstream of the initiation AUG.

4. PROCEDURES FOR PREPARING SINGLE-STRANDED pEMBL DNA

The procedure for preparing a stock of f1 phage for superinfecting bacteria carry-

Table 1. Preparation of Phage f1 Stock.

1.	Inoculate 1 ml of L-broth[a] with a fresh single plaque of f1[b] and let it grow for 3 h at 37°C.
2.	Dilute the culture with 200 ml of L-broth.
3.	Continue to incubate the culture overnight at 37°C.
4.	Pellet the cells by centrifugation and collect the supernatant which contains phage secreted by the infected cells.
5.	Determine the titre of phage in the supernatant by making serial dilutions into L-broth. Add 0.1 ml aliquots of the diluted to 0.2 ml of a saturated culture of *E. coli* 71/18[c]. Add 3 ml of molten top agar kept at 45°C and plate onto L-agar[d]. The plaques may be counted after the plates have been incubated at 37°C for 8 h. The supernatant can be kept at 4°C for 1 − 2 months or frozen at − 20°C in aliquots for a longer period.

[a]The composition of L-broth is Bacto-tryptone 10 g, Bacto-yeast extract 5 g, NaCl 10 g per liter, pH to 7.5 with sodium hydroxide.
[b]We use the variant strain IR1 (3).
[c]The genotype of *E. coli* 71/18 is Δ[*lac-pro*]F' *lac*I^q *lac*ZΔM15 *pro*$^+$ *supE.
[d]L-agar is 15 g Bacto-agar per litre of L broth. TOP-agar is 7 g Bacto-agar per litre of L broth.

Table 2. Preparation of Single-stranded DNA Template.

1.	Inoculate 1 ml of L-broth containing 100 μg/ml of ampicillin with a fresh single colony of cells harbouring pEMBL plasmids and let it grow to saturation at 37°C.
2.	Dilute the culture 1:50 with L-broth and let it grow up to an O.D.$_{660}$ of 0.2.
3.	Infect the cells with phage f1 (IR1) prepared as in *Table 1* at a multiplicity of infection (m.o.i.) of 20 plaque forming units (p.f.u.) per cell[a]. Continue to culture the cells at 37°C for 6 h.
4.	Centrifuge 1 ml of the culture in a microcentrifuge for 5 min.
5.	Remove only 800 μl of the supernatant so as not to disturb the pellet of cells.
6.	Add 200 μl of 2.5 M NaCl, 20% polyethylene glycol (PEG) 6000 to the supernatant and leave at room temperature for 15 min. Then centrifuge the mixture for 5 min in a microcentrifuge.
7.	Remove the supernatant very carefully and resuspend the pellet in 100 μl of 10 mM Tris-HCl pH 8.0, 1 mM EDTA. Extract proteins from the resuspended pellet by shaking gently with 100 μl of phenol saturated with 0.1 M Tris-HCl pH 9.0, 10 mM EDTA.
8.	Centrifuge the mixture for 1 min to separate the phases. Carefully remove the aqueous phase (~ 80 μl) and add 50 μl of isopropanol and 20 μl of 5 M sodium perchlorate.
9.	Centrifuge for 15 min to pellet the nucleic acid precipitate. Remove the supernatant and dissolve the pellet, which is not large enough to be seen in 100 μl of 0.3 M sodium acetate. Add 300 μl of ethanol to the dissolved pellet to re-precipitate the nucleic acids.
10.	Centrifuge the precipitate for 15 min. Carefully remove the supernatant and resuspend the pellet in 30 μl of 10 mM Tris-HCl pH 7.5, 0.1 mM EDTA.

[a]When the culture attains an O.D.$_{660}$ of 0.2, there are approximately 2 x 10^8 cells per ml.

ing pEMBL plasmids is given in *Table 1*. The protocol for superinfecting such cells and the subsequent preparation of s.s. pEMBL DNA is given in *Table 2*. The DNA prepared in this way is suitable for DNA sequencing experiments. The yields obtained are about 0.5 μg of s.s. pEMBL DNA from 1 ml culture. We routinely observe a variability of a factor of 2 − 3 in the total yield of DNA or in the ratio of f1 to plasmid DNA, but we have not been able to identify the factors which influence these yields. In a small percentage of cases colonies which contain a pEMBL type plasmid fail to yield any single-stranded plasmid DNA. We suspect that this is due to some variability between the different colonies since the problem can be overcome by either testing a larger number of colonies or by re-

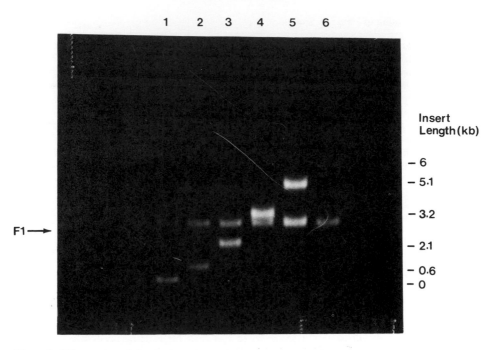

Figure 4. 1% agarose gel electrophoresis of s.s. DNA showing the efficiency of packaging in pEMBL recombinants containing inserts of different sizes.

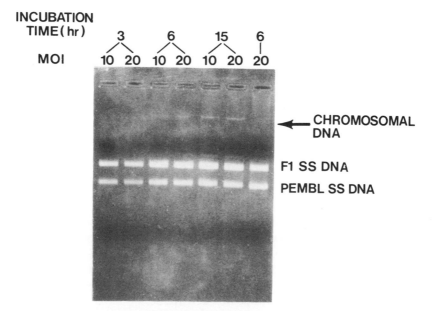

Figure 5. Production of s.s. DNA in different conditions. *E. coli* 71/18 cells containing pEMBL8 were grown up to an $O.D._{660}$ of 0.2 and infected at an M.O.I. of 10 or 20 p.f.u./cell with phage f1 IR1. After different times of incubation at 37°C, s.s. DNA was prepared according to the method described in *Table 2* and analysed by electrophoresis on a 1% agarose gel. A sample prepared from cells infected at an $O.D._{660}$ of 0.4 with a M.O.I. of 20 p.f.u./cell was electrophoresed in lane 6.

introducing plasmids into bacteria by a second round of transformation. The efficiency of production and secretion of s.s. DNA depends on the amount of double-stranded DNA present in the cells at the time of f1 superinfection. We have occasionally observed that foreign DNA inserts interfere with the plasmid replication causing a reduction of copy number. These recombinants will produce considerably less s.s. DNA compared to that of the superinfecting f1. This is clearly the case for the expression vectors that we have described. We have noticed that the efficiency of packaging of plasmids containing strong promoters is inversely correlated with the activity of the promoters. We succeeded in preparing s.s. DNA of the vectors pEMBL ex1 and pEMBL ex2 only in conditions where the λ promoters are completely repressed by an endogenous λ cI gene. The efficiency of packaging is not significantly affected by the size of the plasmid. This is clearly shown in *Figure 4*. Over the wide range of values tested neither the multiplicity of infection nor the concentration of cells at the moment of infection seem to affect the efficiency of packaging (*Figure 5*). Prolonged incubation at 37°C after f1 infection, however, results in contamination of the s.s. DNA by chromosomal DNA. This is probably due to cell death in the late stationary phase.

5. ACKNOWLEDGEMENTS

We thank I.Benner for typing the manuscript. C.Baldari is on leave of absence from the Istituto di Fisiologia Generale, Dipartimento di Genetica e Biologia Molecolare, Universita di Roma, La Sapienza.

6. REFERENCES

1. Messing,J. (1983) *Methods Enzymol.,* **101**, 20.
2. Novander,J., Kempe,T. and Messing,J. (1983) *Gene,* **26**, 101.
3. Dotto,G.P., Enea,V. and Zinder,H.D. (1981) *Virology,* **114**, 463.
4. Dotto,G.P. and Horiuchi,K. (1981) *J. Mol. Biol.,* **153**, 169.
5. Dente,L., Cesareni,G. and Cortese,R. (1983) *Nucleic Acids Res.,* **11**, 1645.
6. Hinnen,A., Hicks,J.B. and Fink,J.R. (1978) *Proc. Natl. Acad. Sci. USA,* **75**, 1929.
7. Struhl,K., Stinchcomb,D.T., Scherer,S. and Davis,R.W. (1979) *Proc. Natl. Acad. Sci. USA,* **76**, 1035.
8. Stinchcomb,D.T., Thomas,M., Kelly,J., Selker,E. and Davis,R. (1980) *Proc. Natl. Acad. Sci. USA,* **77**, 4559.
9. Remaut,H., Stanssens,P. and Fiers,W. (1983) *Nucleic Acids Res.,* **11**, 4677.
10. Queen,C. (1983) *J. Mol. Appl. Genet.,* **2**, 1-10.
11. Rose,M. (1982) Ph.D. thesis, M.I.T.
12. Tschumper,G. and Carlson,J. (1980) *Gene,* **10**, 157.
13. Hartley,J.L. and Donelson,J.E. (1980) *Nature,* **286**, 860.

CHAPTER 6

Techniques for Transformation of *E. coli*

DOUGLAS HANAHAN

1. INTRODUCTION

The ability to transfer plasmids into *E. coli* has come to be an integral part of the repertoire of tools applied in molecular biology. Plasmid transformation of *E. coli* was first demonstrated by Cohen *et al.* (1), applying the observation of Mandel and Higa (2) that the combination of *E. coli* and bacteriophage λ in a solution of CaCl$_2$ at 0°C produced infection (actually transfection). Since then a considerable number of studies have addressed the process of plasmid transformation, with the goals of both improving the frequencies of transformation and characterizing the parameters involved (3 – 15).

The consensus of these investigations is that *E. coli* cells and DNA productively interact in an environment of divalent cations at low temperatures. A variety of factors improve transformation frequencies above those effected by divalent cations alone. These include: brief treatment of the cell/DNA mixture at elevated temperatures (heat shock); the inclusion of monovalent cations in the transformation buffer; the further addition of hexamine cobalt (III) chloride; the treatment of the cells with solvents and sulfhydryl reagents; and growth in medium containing elevated levels of magnesium (10 – 20 mM). These treatments have improved the transformation frequencies from 1/10^5 to as much as 1/10^2 (i.e., one in 100 plasmid molecules effects a transformed cell).

This chapter describes several different techniques for transformation of *E. coli* by plasmids. The methods range from the classic CaCl$_2$ mediated transformation, which is simple but inefficient, to a high frequency method, which is complicated but will produce the very efficient transformations that are necessary for certain classes of experiments, such as cDNA cloning and plasmid rescue. *Table 1* presents a summary of the various protocols described below.

2. CHOOSING A SUITABLE STRAIN

The choice of a particular *E. coli* strain may be dictated by the specific application and perhaps by the transformation protocol being employed. Several general recommendations can be made on strains to be used for standard recombinant DNA experiments. *E. coli* C600 (16) and its many derivatives transform well with all of the protocols described here, and the strain is healthy and easy to work with. MC1061 (17) transforms very well with the 'simple' CaCl$_2$ mediated methods, and enjoys considerable application using such procedures. MC1061 is, however, refractory to efficient transformation by all the procedures

Table 1. Summary of Transformation Protocols.

Table	Procedure	Average transformation frequency[a]	Applicability	Notes
2.	Simple transformation	$10^5 - 10^7$	MC1061[b], C600[c], DH1[d] Most K12 strains	Variation of original Mandel and Higa procedure (2). Simple but inefficient
3.	Standard transformation	$2 - 7 \times 10^8$	DH1, C600, JM109[e], many other K12 strains	Highest transformation frequencies and highest proportion of competent cells (sensitive to contaminants in H_2O and DMSO)
4.	Colony transformation	$0.5 - 2 \times 10^8$	DH1, C600, JM109, many other K12 strains	Fast and convenient, a good general technique (does not require liquid culture)
5.	Rapid transformation	$10^3 - 10^4$	Most K12 strains	Very fast, very inefficient, useful for reintroducing cloned plasmids
6.	Frozen storage I	$10^5 - 10^7$	MC1061, C600, DH1 Most K12 strains	Convenient but inefficient
7.	Frozen storage II	$10^7 - 10^8$	DH1, C600, JM109, many other K12 strains	Convenient and efficient (sensitive to contaminants in H_2O and DMSO)
8.	Frozen storage III	$10^7 - 10^8$	DH1, C600, HB101, JM109, etc.	Convenient and efficient. Does not require DMSO
9.	χ1776 transformation	10^8	χ1776[f]	Only recommended for experiments requiring EKII levels of containment or for transforming cell wall mutants

[a]Transformation frequency is expressed as colonies formed per microgram of pBR322 DNA.
[b]The genotype of *E. coli* MC1061 is F$^-$, ara D139, $\triangle$(ara,leu)7696, $\triangle$lac Y74, gal U$^-$, gal K$^-$, hsr$^-$, hsm$^+$, strA.
[c]The genotype of *E. coli* C600 is F$^-$, supE44, thi-1, leu B6, lac Y1, ton A21, λ^-.
[d]The genotype of *E. coli* DH1 is F$^-$, endA1, hsdR17(rk$^-$,mk$^+$), supE44, thi-1, λ^-, recA1, gyrA96, relA1.
[e]The genotype of *E. coli* JM109 is recA1, endA1, gyrA96, thi, hsdR17, sup E44, relA1, λ^-, $\triangle$(lac-proA,B), F$'$ tra D36, proA,B, lacIqZ-M15.
[f]The genotype of *E. coli* χ1776 is F$^-$ tonA53, dapD8, minA1, glu V44, (sup E42), $\triangle$(gal-uvrB)40, λ^-, minB2, rfb-2, gyr A25, thy A142, oms-2, met C65, oms-1, (tte-1), $\triangle$(bioH-asd)29, cyc B2, hsdR2.

employing dimethylsulphoxide (DMSO), and is therefore not recommended for use with any but the basic CaCl$_2$ mediated methods.

Derivatives of Hoffman Berling strain 1100 (18) have proved very good hosts for recombinant DNA experiments. They are efficiently transformed by all the methods described here. The strains are healthy and produce excellent yields of plasmid DNA. The endonuclease A$^-$ mutation improves the yield and quality of mini-preparations of plasmid DNA. Recommended derivatives of strain 1100 include MM294 (18), DH1 (15), and JM109 (J.Messing, personal communication). JM109 is a new M13 host, which was derived from DH1, and is *recA*$^-$, lacks the

K restriction system, and carries no lysogens. JM109 is thus a significantly better host for M13 based cloning than earlier strains (J.Messing, personal communication). DH1 carries the *rec A1* mutation, and has proved a very good host for many recombinant DNA applications, including cDNA cloning, plasmid rescue, and cloning of large DNA fragments.

Recently, a derivative of DH1, called DH5, has been isolated. This strain transforms significantly better than DH1 with all the protocols listed here, giving transformation efficiencies which range from 2×10^7 transformants/μg pBR322 DNA using the simple protocol to $4-12 \times 10^8$ transformants/μg using the standard high efficiency method. DH5 can be transformed $10-20$ times more efficiently than DH1 by large plasmids (40−60 kb). DH5 remains non-recombinogenic and non-mutagenic.

The *E. coli* Genetic Stock Center carries many of the strains mentioned here, and is an invaluable resource for a wide variety of *E. coli* strains. Requests for strains should be addressed to Dr Barbara Bachmann, ECGSC, Department of Human Genetics, Yale University, 333 Cedar St., P.O. Box 3333, New Haven, CT 06510, USA.

3. MAINTENANCE OF *E. COLI* STRAINS FOR TRANSFORMATION

E. coli is a remarkable resilient organism, and can tolerate a wide variety of growth and storage conditions. However, the preparation of *E. coli* to achieve efficient transformation is relatively sensitive to the manner in which the cells have been stored and cultured immediately before use. Three methods of storing *E. coli* have proved reliable for use in preparing competent cells:

(i)   As a frozen stock in a $-70°C$ freezer or in liquid nitrogen;
(ii)   In slow anaerobic growth in stab bottles;
(iii)   In aerobic growth on agar plates at room temperature.

The time scales of reasonable storage of *E. coli* using these alternatives is $10-20$ years, $1-2$ years, and several days, respectively. The storage of *E. coli* in 50% glycerol at $-20°C$ or at 4°C (on plates or in stabs) have all proved inconsistent, and are not generally recommended.

The three alternative storage methods are described below. In each case, cells should be streaked on an agar plate, and incubated to establish colonies. Several isolated colonies are then picked using a flame sterilized tungsten loop.

3.1 Frozen Stocks

Pick several fresh colonies and inoculate 5 ml of rich medium, such as SOB medium (*Table 12*). Incubate with agitation until the culture is in the mid-to-late phase of logarithmic growth ($1-2 \times 10^8$ cells/ml). Dilute the culture 1:1 in a solution of 60% growth medium, 40% glycerol. Aliquot 1 ml volumes into a series of 2 ml screw cap polypropylene tubes (or similar containers), chill on ice, flash freeze in liquid nitrogen, and place at $-70°C$. To retrieve cells, remove a tube from the freezer, immediately scrape a chunk of frozen cells off the top of the frozen mass, and return the tube to the freezer. Attempt to minimize warming and thawing the contents of the tube. Touch the frozen clump of cells to an agar plate, where it will immediately melt. Spread the drop to isolate single cells, and incubate the plate to establish colonies. It is good policy to prepare a set of serially numbered tubes of frozen cells, and then to retrieve cells from each tube 10 – 30 times, before discarding it and going on to the next. (The titer of viable cells declines with the repeated brief removals from the freezer. When only a few cells are produced from a clump, that tube should be discarded.)

3.2 Stabs

Prepare a solution of growth medium such as SOB (*Table 12*) containing 0.8% Bacto Agar. Autoclave or place briefly in a microwave oven to dissolve the agar, and dispense 4 ml volumes into 5 ml glass Bijou bottles (or similar small bottles that can be tightly sealed). Autoclave to sterilize, and then leave the bottles loosely capped overnight to allow evaporation of condensed water. Pick several fresh, isolated colonies using a flame sterilized tungsten loop. Flame the cap of the stab bottle. Remove the cap, quickly stab the loop into the agar, and close the bottle tightly. Store the stab in the dark at room temperature (not at 4°C).

3.3 Plates at Room Temperature

Streak cells on an agar plate, and incubate at 37°C to establish colonies 1 – 2 mm in diameter. Store at room temperature. If old, fat colonies are being used to inoculate a culture or to restreak a fresh plate, it is best to pick the edge of the colony, where the cells are still growing.

4. TRANSFORMATION PROCEDURES

4.1 Simple Transformation

The basic outlines of the treatments employed by Mandel and Higa have endured in mostly unaltered form, in part because of their inherent simplicity and general applicability to a wide variety of strains of *E. coli*. The protocol described in *Table 2* has several minor modifications, which seem to improve the transformation frequency and reproducibility.

The conditions specified here represent a good compromise for a variety of *E. coli* strains. If a particular strain shows poor transformation frequencies, a few simple modifications may prove beneficial. The first is to grow the cells in medium containing 10 – 20 mM $MgCl_2$, a condition which is very important for the complex transformation protocols described below, but is of variable benefit to this protocol. For example, addition of Mg^{2+} to the growth medium improves transformation of DH1, and reduces that of MC1061. This likely reflects differences in the constitution of the respective cell envelopes. A second type of modification is to substitute transformation buffers, in particular to use either TFB or SB2 (which contains $MnCl_2$ plus $CaCl_2$, see *Table 10*). Either of these alternative buffers may prove beneficial with strains that are not effectively transformed by the simple protocol. Thus, one or two days of work spent in comparing transformation buffers may significantly improve the transformation of a strain which has been chosen for genetic reasons rather than for demonstrating proficiency in plasmid transformation.

Table 2. Protocol for Simple Transformation.

1.	Pick several 2–3 mm diameter colonies off a freshly streaked plate and disperse the cells by vortexing in 1 ml of growth medium[a]. Use a flame sterilized tungsten inoculating loop, and avoid scraping up pieces of agar.
2.	Inoculate the dispersed cells into an Erlenmeyer flask containing growth medium. A good rule is to use one colony per 10 ml of growth medium, and to use a volume of medium that comprises less than 10% of the flask volume. SOB medium[b] (either with or without added Mg^{2+}, depending on the strain) is recommended.
3.	Incubate the culture at 37°C with moderate agitation[c] until the culture is midway through the logarithmic phase of growth and contains about 4–9 x 10^7 viable cells/ml[d].
4.	Collect the culture into centrifuge tubes such as 50 ml polypropylene Falcon 2070 tubes or their equivalent and chill on ice for 10–60 min.
5.	Pellet the cells by centrifugation at 750–1000 g (~2000–3000 r.p.m. in a standard clinical centrifuge) for 15 min at 4°C. Decant the supernatant, and briefly leave the tube inverted on paper towels, rapping the tube if necessary to remove remaining liquid.
6.	Disperse the cells in 1/3 of the culture volume of simple transformation buffer (SB)[e] (or standard TFB[e])by either vortexing moderately or by pipetting the cells up and down. Incubate on ice for 10–60 min.
7.	Pellet the cells by centrifugation for 15 min at 750–1000 g. Decant the supernatant, draining thoroughly as in step 5.
8.	Resuspend the cells in a volume of SB or TFB that is 1/12.5 of the original culture volume, or the fraction thereof placed in each centrifuge tube. Thus, 2.5 ml of culture are concentrated into 200 μl of SB, which represents the standard unit transformation volume.
9.	Pipette 200 μl aliquots of the cells into separate tubes (e.g., 17 x 100 mm or 12 x 60 mm polypropylene, or standard glass test tubes).
10.	Add the DNA solution in volume of <20 μl. Swirl the tube to mix the DNA evenly with the cells. Incubate on ice for 10–60 min.
11.	Heat shock the cells by placing the tubes in a 42°C water bath for 90 sec and then chill by returning the tubes immediately to 0°C (crushed ice).
12.	Add 800 μl of growth medium and incubate at 37°C with moderate agitation for 30–60 min[f]. SOC medium[b] is recommended.

[a]When transforming MC1061 use liquid and solid medium that lacks Mg^{2+}. For DH1 and C600 strains, include 20 mM Mg^{2+} in the medium. See text for further details.
[b]See *Table 12*.
[c]Moderate agitation is given by shaking on a rotatory shaker with a 'throw' of 1–2 cm at a speed of 250 r.p.m., or under equivalent conditions.
[d]4–9 x 10^7 cells/ml corresponds to a culture of *rec*[−] strains at O.D.$_{550}$ of 0.35–0.60, and to cultures of *rec*[+] strains at O.D.$_{550}$ of 0.2–0.4.
[e]See *Table 10*.
[f]Omit this incubation for M13 transfections.

The protocol in *Table 2* is easy, but it has certain limitations. In particular, the transformation frequency, which is generally scored as colonies formed per microgram of pBR322, is low (ranging from 10^5 to 10^7), as is the fraction of competent cells (usually 10^{-3} to 10^{-6}). This second figure affects the colony forming potential of a preparation of competent cells. The manifestation of this effect is a saturation curve, producing increasingly fewer additional colonies with each incremental addition of DNA. As a general rule, add less than 10 nanograms of transforming plasmid DNA to a preparation of cells. This figure may, of course, be different from the actual quantity of plasmid DNA in a ligation or annealing solution, where a significant fraction of the plasmid DNA may be nontransforming (often linear) DNA that has not been recombined with insert to produce a circular molecule that will then efficiently transform.

4.2 Standard High Efficiency Transformation

The 'standard' transformation procedure described in *Table 3* is a recapitulation of that which evolved from a systematic investigation of factors influencing plasmid transformation of *E. coli* (15). This method has the advantage that it produces very high transformation frequencies; it gives efficient transformation with both large and relaxed plasmids; and it produces a high proportion of cells competent for transformation. The preparation of competent cells using this method requires considerable attention to detail. The efficiency of the method suffers from sensitivity to a variety of contaminants. When one is familiar with the details of the preparation of solutions and when the purity of reagents has been ascertained, it is possible to obtain routinely the high transformation frequencies necessary for many experiments. The transfection frequency of the replicative form of bacteriophage M13 is comparable with the transformation frequency of a similarly sized plasmid, and thus this procedure will efficiently introduce M13 vectors into appropriate host cells.

The protocol in *Table 3* contains one modification of the original (15). The DMSO and dithiothreitol (DTT) have been combined, such that the three separate additions have now become two. This simplification produces transformation frequencies equal to or greater than those achieved with the separate additions using the *E. coli* strain DH1 and its relatives. Furthermore, the DMSO + DTT solution may be added in one aliquot, with frequencies of about 75% of the values obtained with two additions. The original version is summarized at the bottom of the table.

The preparation of the culture of *E. coli* to be used in a transformation is of considerable importance when reproducibility and reliability are desirable, that is most of the times one would employ this procedure. Many growth conditions will produce a population of cells that transform well, but the following ritual has proved most consistent. A clump of cells is removed from a stock frozen at $-70°C$ and streaked on a SOB plate (*Table 12*), and incubated at $37°C$ to establish colonies $2.5-3$ mm in diameter. Colonies are picked and dispersed in 1 ml of SOB medium, inoculated into a larger volume of that medium, and the culture incubated with agitation at $37°C$ until the cell density reaches $5-7 \times 10^7$ viable cells per ml. The culture is then collected into chilled centrifuge tubes and carried through the various steps of the transformation. Other methods of storage and preparation will suffice, but none have proved as reliable in the hands of the author. Fresh stabs, restreaks of growing cultures and cells growing slowly at room temperature have all been satisfactory and may be considered as alternatives, bearing in mind their possible inconsistencies. In contrast, cells derived from plates or stabs stored at $4°C$, or from cell suspensions kept in 50% glycerol at $-20°C$, have not been generally reliable. The physiological basis for the observed irregularity in transformation, given independent cultures collected at the same cell density, remains uncertain. The routine outlined above satisfies the requirements for consistency though not the desire for an understanding of the

Table 3. Standard Transformation Protocol.

1.	Pick several 2 – 3 mm diameter colonies off a freshly streaked SOB[a] agar plate and disperse in 1 ml SOB medium by vortexing. Use one colony per 10 ml of culture medium. The cells are best streaked from a frozen stock or a fresh stab about 16 – 20 h prior to initiating liquid growth.
2.	Inoculate the cells into an Erlenmeyer flask containing SOB medium. Use a culture volume to flask volume of between 1:10 and 1:30 (e.g., 30 – 100 ml in a 1 l flask).
3.	Incubate at 37°C with moderate agitation[b] until the cell density is 4 – 7 x 10^7 viable cells/ml[c].
4.	Collect the culture into 50 ml polypropylene centrifuge tubes such as Falcon 2070 tubes and chill on ice for 10 – 15 min.
5.	Pellet the cells by centrifugation at 750 – 1000 g (2000 – 3000 r.p.m. in a clinical centrifuge) for 12 – 15 min at 4°C. Drain the pelleted cells thoroughly, by inverting the tubes on paper towels and rapping sharply to remove any liquid. A micropipette can be used to draw off recalcitrant drops.
6.	Resuspend the cells in 1/3 of the culture volume of TFB[d] by vortexing moderately. Incubate on ice for 10 – 15 min.
7.	Pellet the cells and drain thoroughly as in step 5.
8.	Resuspend the cells in TFB to 1/12.5 of the original volume. This represents a concentration from 2.5 ml of culture into 200 μl of TFB.
9.	Add DMSO and DTT solution (DnD)[d] to 3.5% (v/v). That is to say, add 7μl per 200 μl of cell suspension. Squirt the DnD into the centre of the cell suspension and immediately swirl the tube for several seconds. Incubate the tubes on ice for 10 min.
10.	Add a second equal aliquot of DMSO and DTT (DnD) as in step 9 to give a 7% final concentration. Incubate the tubes on ice for 10 – 20 min.
11.	Pipette 210 μl aliquots into chilled 17 x 100 mm polypropylene tubes (Falcon 2059 or their equivalent).
12.	Add the DNA solution in a volume of <20 μl, swirling to mix. Incubate the tubes on ice for 20 – 40 min.
13.	Heat shock the cells by placing tubes in a 42°C water bath for 90 sec. Return tubes into ice to quench the heat shock, allowing 2 min for cooling.
14.	Add 800 μl of SOC medium[a] to each tube. Incubate at 37°C with moderate agitation for 30 – 60 min[e]. Spread the cells on agar plates containing appropriate additives to select transformants.

Variations

(i)   DMSO and DTT may be added separately. In this case the following steps apply:

9a. Add DMSO to a concentration of 3.5% (v/v) (i.e., by adding 7 μl per 200 μl cell suspension) by squirting it into the centre of the cell suspension and immediately swirling the tube for several seconds. Incubate the tube on ice for 10 min.

9b. Add DTT solution to 3.5% (v/v) (i.e., by adding 7 μl of 2.2 M DTT, 10 mM KAc pH 6.2 per 200 μl cell suspension) and swirl the tube for 5 sec. Incubate the tube on ice for 10 min.

10. Add a second equal aliquot of DMSO, as before, giving a 7% final concentration. Incubate on ice for 5 – 10 min.

(ii)   The DMSO + DTT solution may be added as one aliquot of 7% (v/v) which is 14 μl per discrete 200 μl aliquot of transformation cocktail. (The transformation frequency is ~75% of that obtained with two separate additions of 3.5%).

---

[a]See *Table 12*.
[b]See footnote[c] to *Table 2*.
[c]For DH1 this corresponds to an O.D.$_{550}$ of 0.45 – 0.55.
[d]See *Table 10*.
[e]This step should be omitted for M13 transfections.

process.

Aside from preparation of the culture of *E. coli*, the other important details affecting transformation efficiency involve the purity of the materials employed. In contrast, most of the incubation times, centrifugation forces and times, and exact concentrations of the components are all relatively flexible. The primary impurities that affect transformation appear to involve organic compounds in the water and oxidation products in the DMSO. Given supplies of good water and DMSO, efficient transformations can be routine. Assessment of these reagents is discussed in the section on materials. A third source of problems relates to contamination by plastics and soap. Glassware used in transformations can be shown to adversely affect the preparation of competent cells. The likely causes are organic compounds in the autoclave produced by sterilizing plastic garbage and soap not completely removed during washing. One solution is to set aside flasks to be used in transformations. These flasks are rinsed thoroughly after use, autoclaved half full of water, and then rinsed with sterile water or medium prior to subsequent use. In addition, some sterile filtration membranes (e.g., Nalge) contain wetting agents such as the detergent Triton X-100. These filters should be rinsed once or twice with water before being used to sterilize TFB or culture medium by filtration.

A fourth impurity relates to surfactants left on plasticware during its manufacture. Certain brands of polypropylene tubes (e.g., Sarsted) must be well rinsed before use — otherwise poor transformations will result. Polystyrene tubes cannot be used with this method, as DMSO will slightly dissolve the tubes, and the dissolved material significantly inhibits transformation. Glass test tubes may be used in place of polypropylene tubes for the transformation reactions. However, the heat shock time should be calibrated for the particular tube in use, as the thermal conductivity of glass differs from that of polypropylene, and there is variability in the wall thickness of different brands of test tubes.

A final source of problems relates to the chemicals and extracts used in growth and preparation of the cells. Digests of casein and extracts of autolyzed yeast are generally equivalent, but problems have been observed with occasional lots of these products. In particular, one brand of yeast extract (Oxoid) has proved inappropriate for preparing competent cells. The chemicals are normally quite stable, but not above contamination or degradation with age. MES, for example, may degrade upon extended storage as a solid at room temperature. Such problems can only be uncovered by substitution of alternative stocks of each compound. The standard transformation buffer originally used RbCl as the monovalent cation. However, KCl has proved equivalent to RbCl (15), and is considerably less expensive, and is now therefore used in formulation of standard TFB. It is recommended that ultrapure KCl be used, due to the high concentration of KCl in this buffer. TFB is very stable, and buffer stored for 2 years at 4°C has proved indistinguishable from a freshly prepared solution.

The transformation frequency obtained using the standard transformation protocol becomes non-linear in the range of $1-5$ ng of closed circular pBR322, and all the competent cells in the population are saturated for their transformation potential at $100-200$ ng of plasmid DNA. Note in addition that the use of plasmid levels above 25 ng produces an appreciable frequency of double trans-formants (15). Thus, it is recommended that the amount of closed circular (trans-formable) plasmid DNA that is added to a discrete transformation mixture be kept below 10 ng. In cases where a significant fraction of plasmid DNA is non-transforming, the level of DNA may be raised, bearing in mind that any DNA becomes competitive in the range of 25 to 100 ng for transformation of 10^8 cells.

4.3 Colony Transformation

The colony transformation protocols are simplifications of the complex proced-ure described above, and are based on the observation that colonies of fresh cells can be picked, dispersed in transformation buffer (TFB, *Table 10*), and carried through the remainder of the protocol. This method avoids the growth of *E. coli* in liquid culture, and is thus advantageous for certain applications. There are two versions of the protocol, which are dubbed colony transformation (*Table 4*) and rapid transformation (*Table 5*).

The colony transformation protocol described in *Table 4* produces transform-ation efficiencies ranging from 1 x 10^7 to 2 x 10^8 colonies/μg pBR. It requires on-ly a plate of relatively fresh cells. This method is convenient and is recommended for subcloning and plasmid constructions. Cells can be streaked on a rich plate containing $10-20$ mM Mg^{2+}, incubated overnight at 37°C to establish colonies, and then further incubated at room temperature until the procedure is initiated

Table 4. Colony Transformation Procedure.

1.	Pick $2-4$ colonies of freshly streaked bacteria[a] from a plate containing $10-20$ mM Mg^{2+} in the medium. Disperse the colonies by vigorous vortexing in 200 μl TFB, in a 17 x 100 mm polypropylene tube (e.g., a Falcon 2059 tube or its equivalent). Remove the majority of the colony mass, but avoid scraping up pieces of agar. The use of a tungsten inoculating loop is recommended.
2.	Incubate the tube on ice for $15-30$ min.
3.	Add 7 μl DMSO and DTT solution (DnD)[b] into centre of solution, swirling for several seconds. Leave the tube on ice for $10-15$ min.
4.	Add another 7 μl of DND, swirl, and incubate on ice for $10-20$ min.
5.	Add the DNA solution in <20 μl. Incubate on ice for $10-40$ min.
6.	Heat shock the cells by placing the tubes in a 42°C water bath for 90 sec, and then chill on ice for 2 min.
7.	Add 800 μl SOC[c] medium. Incubate[d] at 37°C with moderate agitation[e] for $30-60$ min. Plate the bacteria on appropriate selective media.

Variation

DMSO and DTT may be added separately, as outlined in *Table 3*.
The DMSO + DTT solution may also be added as one 14 μl aliquot.

[a]The colonies should be $2-4$ mm in diameter, and are best picked from plates at room temperature. The plate may either be transferred from 37°C to 20°C for >1/2 h, or incubated at 20°C for up to 3 days following initial incubation at 37°C. The cell suspension should be clearly visible following the dispersal of the colonies, otherwise add additional colonies until a thick suspension is produced.
[b]See *Table 10*.
[c]See *Table 12*.
[d]This step should be omitted for M13 transfections.
[e]See footnote[c] to *Table 2*.

Table 5. Rapid Transformation.

1.	Pick several colonies (or a clump of cells) off a plate using a tungsten inoculating loop, being careful to take no agar along with the cells.
2.	Disperse the colonies in 200 μl of standard transformation buffer (TFB[a]) by vigorous vortexing, or by repeated pipetting.
3.	Incubate the cells on ice for 10 min.
4.	Add DNA solution (10−200 ng) in (<20 μl), swirl to mix, and incubate on ice for 10 min.
5.	Heat shock the cells at 37−42°C for 90 sec.
6.	Plate several fractions (e.g., 1%, 5%, 25%) on appropriate selective medium.

[a]See *Table 10.*

(up to 2 days). There are two principal requirements for achieving reliable transformations. The colonies should be 3−5 mm in diameter, and four to two colonies (respectively) should be taken for each discrete 200 μl transformation. In addition, it is important that no agar be taken along with the colonies, as agar proves inhibitory to transformation. A tungsten inoculating loop, such as a Jorgenson inoculating loop, Scientific Products #N2010, or its equivalent works very well for this purpose. When sufficient colony mass is taken with no agar fragments, reliable transformations will be produced. The principal limitation of this method for preparing competent cells is that the saturation levels are reduced relative to a similar population prepared from liquid culture. In other words, the fraction of competent cells in the population is reduced, even though the transformation frequencies achieved are comparable with those obtained with liquid culture. Thus the method is not recommended when the goal is to maximize the total colony forming potential of a sample of DNA, as would be the case for cDNA cloning and plasmid rescue.

The rapid transformation protocol described in *Table 5* is a simplification of the colony transformation protocol in which cells are dispersed in TFB, DNA is added, and the cells are plated (perhaps after a heat shock). This is a very rapid and very inefficient procedure ($\sim 10^3 - 10^4$ colonies/μg). Its principal merit is in rapidly reintroducing cloned plasmids into *E. coli*. One alternative to maintaining recombinant plasmids in *E. coli* cells (as stabs or frozen stocks) is to store the plasmids as DNA. When a particular plasmid is required for some purpose, it can be rapidly reintroduced into *E. coli* and thereby amplified during culture. This eliminates storing and passaging large numbers of transformed strains, and further protects against loss of a strain due to death or contamination, as well as allowing more efficient use of storage space. Where a laboratory keeps hundreds of recombinant plasmids, their maintenance and long-term storage becomes a significant burden, and the author finds that long-term storage of plasmids as DNA is reliable and convenient. Only nontransformed strains of *E. coli* are kept as frozen stocks, and transformed strains in frequent use are kept as stabs, while the remainder of plasmids are kept as DNA at −20°C.

4.4 Frozen Storage of Competent Cells

E. coli that has been rendered competent for transformation can be stored for extended periods at or below −70°C in that state. There are several adaptations

Table 6. Frozen Storage of Competent Cells (Protocol 1).

A. Preparation

1. Pick several 2−3 mm diameter colonies off a freshly streaked plate and disperse the cells by vortexing in 1 ml of growth medium. Use a flame sterilized tungsten inoculating loop, and avoid scraping up pieces of agar.
2. Inoculate the dispersed cells into an Erlenmeyer flask containing growth medium. A good rule is to use one colony per 10 ml of growth medium, and to use a volume of medium that comprises less than 10% of the flask volume. SOB medium (either with or without added Mg^{2+}, depending on the strain) is recommended.
3. Incubate the culture at 37°C with moderate agitation until the culture is midway through the logarithmic phase of growth and contains about $4−9 \times 10^7$ viable cells/ml.
4. Collect the culture into centrifuge tubes such as 50 ml polypropylene Falcon 2070 tubes or their equivalent and chill on ice for 10−60 min.
5. Pellet the cells by centrifugation at 750−1000 g (~2000−3000 r.p.m. in a standard clinical centrifuge) for 15 min at 4°C. Decant the supernatant, and briefly leave the tube inverted on paper towels, rapping the tube if necessary to remove remaining liquid.
6. Disperse the cells in 1/3 of the culture volume of simple frozen storage buffer (FB)[a] by either vortexing moderately or by pipetting the cells up and down. Incubate on ice for 10−60 min.
7. Pellet the cells by centrifugation for 15 min at 750−1000 g. Decant the supernatant, draining thoroughly as in step 5.
8. Resuspend the cells in a volume of FB that is 1/12.5 of the original culture volume, or the fraction thereof placed in each centrifuge tube. Thus 2.5 ml of culture are concentrated into 200 μl of FB, which represents the standard unit transformation volume.
9. Pipette 200 μl aliquots of cell suspension into 2 ml screw cap polypropylene tubes, 1.5 ml microcentrifuge tubes, or snap cap polypropylene tubes.
10. Flash freeze by placing the tubes in a solid CO_2/alcohol bath or in liquid nitrogen for several minutes.
11. Place the tubes at −70°C or in liquid nitrogen.

B. Use of Frozen Competent Cells

1. Remove tube(s) from the freezer and thaw in air at room temperature until the cell suspension is just liquid. Place the tubes on ice. (Aliquot 200 μl volumes to separate tubes if multiples were combined for storage.)
2. Add the DNA solution in a volume of <20 μl. Swirl the tube to mix the DNA evenly with the cells.
3. Incubate the tube(s) on ice for 10−60 min.
4. Heat shock the cells by placing the tubes in a 42°C water bath for 90 sec, and then chill by returning tubes immediately to 0°C (crushed ice).
5. Add 800 μl of SOC medium[b] and incubate at 37°C with moderate agitation for 30−60 min[c].

[a]See *Table 10* for the composition of FB buffer.
[b]See *Table 12* for the composition of SOC medium.
[c]Omit this step for M13 transfections.

of the transformation procedures that can be used to prepare and store frozen competent cells. All have the advantage that large numbers of competent cells can be prepared on one occasion, and then aliquots used for transformation for many months. Transformation frequencies decline slowly over time. The fractions of viable and of competent cells in a frozen and thawed preparation are generally reduced relative to a freshly prepared one. Thus frozen competent cells are not recommended when the highest frequencies and greatest colony forming potential are required, such as when establishing cDNA libraries from limiting amounts of DNA.

Table 7. Frozen Storage of Competent Cells (Protocol 2).

A. Preparation

1. Pick several 2 – 3 mm diameter colonies off a freshly streaked SOB agar plate and disperse in 1 ml SOB medium by vortexing. Use one colony per 10 ml of culture medium. The cells are best streaked from a frozen stock or a fresh stab about 16 – 20 h prior to initiating liquid growth.

2. Inoculate the cells into an Erlenmeyer flask containing SOB medium. Use a culture volume to flask volume of between 1:10 and 1:30 (e.g., 30 – 100 ml in a 1 l flask).

3. Incubate at 37°C with moderate agitation until the cell density is 6 – 9 x 10^7 viable cells/ml.

4. Collect the culture into 50 ml polypropylene centrifuge tubes such as Falcon 2070 tubes and chill on ice for 10 – 15 min.

5. Pellet the cells by centrifugation at 750 – 1000 g (2000 – 3000 r.p.m. in a clinical centrifuge) for 12 – 15 min at 4°C. Drain the pelleted cells thoroughly, by inverting the tubes on paper towels, and rapping sharply to remove any liquid. A micropipette can be used to draw off recalcitrant drops.

6. Resuspend the cells in 1/3 of the culture volume of frozen storage buffer (FSB)[a] by vortexing moderately. Incubate on ice for 10 – 15 min.

7. Pellet the cells as before, and drain thoroughly as in step 5.

8. Resuspend the cells in FSB to 1/12.5 of the original volume. This represents a concentration from 2.5 ml of culture into 200 µl of FSB.

9. Add DMSO to 3.5% (v/v). This corresponds to 7 µl per 200 µl of cell suspension. Squirt the DMSO into the centre of the cell suspension and immediately swirl the tube for 5 – 10 sec. Incubate the tube on ice for 5 min.

10. Add a second equal aliquot of DMSO, as before, giving a 7% final concentration. Incubate on ice for 10 – 15 min. (DTT is not used in this protocol.)

11. Pipette 210 µl aliquots into chilled screw cap polypropylene tubes, 1.5 ml microcentrifuge tubes, or into snap cap polypropylene tubes.

12. Flash freeze by placing the tubes in a dry ice/alcohol bath or into liquid nitrogen for several minutes.

13. Transfer tubes to a − 70°C freezer or into liquid nitrogen.

B. Use of Frozen Competent Cells

1. Remove tube(s) from the freezer and thaw in air at room temperature until the cell suspension is just liquid. Place the tubes on ice. (Aliquot 200 µl volumes to separate tubes if multiples were combined for storage.)

2. Add the DNA solution in a volume of < 20 µl. Swirl the tube to mix the DNA evenly with the cells.

3. Incubate the tube(s) on ice for 10 – 60 min.

4. Heat shock the cells by placing the tubes in a 42°C water bath for 90 sec, and then chill by returning tubes immediately to 0°C (crushed ice).

5. Add 800 µl of SOC medium[b] and incubate at 37°C with moderate agitation for 30 – 60 min[c].

Three methods for preparing frozen competent cells are presented in *Tables 6* *to 8*. All employ glycerol as a stabilizer. The first is a variation on the standard CaCl$_2$ mediated transformation procedure. It is the simplest to prepare, and is also the most inefficient with most strains. However, certain strains (e.g., MC1061) transform better with this than with those employing more complex buffers.

The second protocol is an adaptation of the standard transformation protocol to prepare frozen competent cells for storage. Glycerol has been added, and potassium acetate substituted for 2[N-morphino]ethone sulphonic acid (MES). DTT and MES are inhibitory in the preparation and use of frozen competent cells, and thus both are omitted in this procedure. The optimum cell density for preparing frozen cells seems to be slightly higher than for immediate use. In the case of DH1, cells are collected at 6 – 9 x 10^7 viable cells/ml rather than 5 – 7 x 10^7 for

Table 8. Frozen Storage of Competent Cells (Protocol 3).

A. Preparation

1. Pick several 2 – 3 mm diameter colonies off a freshly streaked SOB agar plate and disperse in 1 ml SOB medium by vortexing. Use one colony per 10 ml of culture medium. The cells are best streaked from a frozen stock or a fresh stab about 16 – 20 h prior to initiating liquid growth.
2. Inoculate the cells into an Erlenmeyer flask containing SOB or Psi[a] medium. Use a culture volume to flask volume of between 1:10 and 1:30 (e.g., 30 – 100 ml in a 1 l flask).
3. Incubate at 37°C with moderate agitation until the cell density is 4 – 7 x 10^7 viable cells/ml.
4. Collect the culture into 50 ml polypropylene centrifuge tubes such as Falcon 2070 tubes and chill on ice for 10 – 15 min.
5. Pellet the cells by centrifugation at 750 – 1000 g (2000 – 3000 r.p.m. in a clinical centrifuge) for 12 – 15 min at 4°C. Drain the pelleted cells thoroughly by inverting the tubes on paper towels, and rapping to remove any liquid. A micropipette can be used to draw off recalcitrant drops.
6. Resuspend the cell pellet by moderate vortexing in a volume of *RF1*[b] that is 1/3 of the volume collected. Incubate the cells on ice for 15 min (for DH1 and JM101) to 2 h (for HB101).
7. Pellet cells as in step 5 of *Table 3*.
8. Resuspend the cells in *RF2* to 1/12.5 of the original volume. Incubate the cells on ice for 15 min.
9. Distribute aliquots into chilled screw cap tubes or 1.5 ml microcentrifuge tubes.
10. Flash freeze in a solid CO$_2$/alcohol bath or in liquid nitrogen, then place at – 70°C.

B. Use of Frozen Competent Cells

1. Remove tube(s) from the freezer and thaw in air at room temperature until the cell suspension is just liquid. Place the tubes on ice. (Aliquot 200 μl volumes to separate tubes if multiples were combined for storage.)
2. Add the DNA solution in a volume of <20 μl. Swirl the tube to mix the DNA evenly with the cells.
3. Incubate the tube(s) on ice for 10 – 60 min.
4. Heat shock the cells by placing the tubes in a 42°C water bath for 90 sec, and then chill by returning tubes immediately to 0°C (crushed ice).
5. Add 800 μl of SOC medium[b] and incubate at 37°C with moderate agitation for 30 – 60 min[c].

[a]See *Table 12* for the composition of media.
[b]See *Table 10* for the composition of *RF1* and *RF2*.

the standard protocol.

The third procedure (*Table 8*) was developed by Viesturs Simanis (Imperial College, London; personal communication). This protocol produces competent cells with average transformation frequencies similar to those of *Table 7*, but has the advantage that it does not require DMSO. Thus, if DMSO is suspected to be inadequate, this procedure will produce efficient transformations in its absence. Again, certain strains may transform better with this procedure than with those listed above.

4.5 Transformation of χ1776

The method described here was developed during the period when virtually all recombinant DNA experiments required the use of highly enfeebled bacterial hosts. The *E. coli* strain χ1776 was constructed by Curtiss and his collaborators as a host for recombinant plasmids (19), and was certified as an EKII host due to its high sensitivity to soap and bile salts and its unusual auxotrophic requirements. The procedure in *Table 9* produces transformation frequencies of $1-2$ x 10^8 transformants/μg pBR322 using χ1776, with $10-12\%$ of the cells in a preparation being competent for transformation. It is presented here for several reasons. First, χ1776 remains a certified EKII host which may be deemed necessary for performing certain classes of experiments, such as cloning the genes for toxins. Second, χ1776 carries a number of alterations in its cell envelope. It is possible that other strains of *E. coli* carrying some of these alterations will prove amenable to efficient transformation by this method. Thus, if a strain does not adequately transform with the protocols described above, the χ1776 protocol should be tried.

Table 9. Protocol for Transformation of χ1776.

1.	Pick several $2-3$ mm diameter colonies off a freshly streaked Xb[a] plate. Disperse in 1 ml of Xb medium by gently vortexing.
2.	Inoculate $25-100$ ml Xb in a 300 ml flask using 1 colony per 10 ml culture volume. Each discrete transformation requires 5 ml of culture.
3.	Incubate the culture at 37°C with vigorous agitation until the cells reach a density of $5-7$ x 10^7 viable cells/μl[b]. The last two doublings occur in $38-45$ min if the cells are growing well.
4.	Collect the culture into 50 ml polypropylene centrifuge tubes such as Falcon 2070 tubes or their equivalent and chill on ice for $10-15$ min.
5.	Pellet the cells by centrifugation at $750-1000$ g for $12-15$ min at 4°C.
6.	Drain the pellet well by inverting the tubes on a pad of paper towels, and rapping to remove drops of liquid.
7.	Resuspend the pellet in 1/3 of the culture volume of XFB[a]. Swirl or vortex gently to disperse the pellet. Incubate on ice for $5-10$ min.
8.	Pellet the cells as in steps 5 and 6.
9.	Resuspend the cells in 1/25 of the culture volume of XFB. This concentrates each 5 ml of culture into 200 μl, an amount sufficient for one discrete transformation. Incubate on ice for $10-15$ min.
10.	Add 7 μl of DMSO per 200 μl cells, squirting it directly into the centre of the cell suspension, and swirling immediately. Incubate the cells on ice for $5-25$ min, swirling occasionally[c].
11.	Pipette 200 μl aliquots into 12 x 75 mm polypropylene tubes such as Falcon 2063 tubes or their equivalent.
12.	Add DNA in a volume of $1-10$ μl. Incubate on ice for $10-30$ min.
13.	Freeze shock the cells by placing the tubes in a -70°C bath for 100 sec. The suspension should freeze solidly. A dry ice/isopropanol bath is appropriate. Remove tubes from the cold bath and thaw them in air at room temperature until the suspension has just reached the liquid state[d].
14.	Incubate cells on ice for $5-10$ min.
15.	Heat shock the cells by placing the tubes in a 42°C water bath for 60 sec. If tube dimensions or material are different, the optimal times for heat shock may be different as well. Place tubes on ice for 2 min.
16.	Add 800 μl Xb, and incubate at 37°C for 1 h with *no agitation*. Spread on Xb plates with the appropriate selective system.

[a]See *Table 12* for the composition of Xb medium.
[b]This corresponds to an O.D.$_{550}$ of 0.2.
[c]The final concentration of DMSO in this protocol is 3.5% (v/v). There is only one addition of DMSO, and DTT is not used, in contrast to the method in *Table 3*.
[d]Frozen competent cells of strain χ1776 may be stored for extended periods at -70°C. In this case omit the addition of DNA (step 12) prior to the freeze shock step, and instead place the frozen cells at -70°C. Add the DNA after the cells have been thawed and placed on ice.

The protocol described here is quite similar to the standard transformation method, and in fact was a predecessor to it (15). A major difference is the use of a freeze shock step, in which the cells are rapidly frozen and thawed while suspended in transformation buffer. This freeze shock has proved effective only with $\chi1776$, and is thus likely to relate to the specific architecture of its cell envelope.

Table 10. Formulations of Transformation Buffers.

A. *Standard Transformation Buffer (TFB)*

Compound	Amount/litre	Final concentration
KCl (ultrapure)	7.4 g	100 mM
MnCl.4H$_2$O	8.9 g	45 mM
CaCl$_2$.2H$_2$O	1.5 g	10 mM
HACoCl$_3$[a]	0.8 g	3 mM
K-MES	20 ml of 0.5 M stock (pH 6.3)	10 mM
(final pH 6.20 ± 0.10)		

Preparation: Equilibrate a 0.5 M solution of MES (2[N-morpholino]ethone sulphonic acid) to pH 6.3 using concentrated KOH, then sterilise by filtration through a 0.22 μ membrane, and store in aliquots at $-20°C$. Make a solution of 10 mM K-MES, using the 0.5 M MES stock and the purest available water. Add the salts as solids, then filter the solution through a 0.22 μ pre-rinsed membrane. Aliquot into sterile flasks, and store at 4°C. (TFB is stable for >1 year.)

B. *Frozen Storage Buffer (FSB)*

Compound	Amount/litre	Final concentration
KCl	7.4 g	100 mM
MnCl$_2$.4H$_2$O	8.9 g	45 mM
CaCl$_2$.2H$_2$O	1.5 g	10 mM
HACoCl$_3$[a]	0.8 g	3 mM
Potassium acetate	10 ml of a 1 M stock (pH 7.5)	10 mM
Redistilled glycerol	100 g	10% (w/v)
(final pH 6.20 ± 0.10)		

Preparation: Equilibrate a 1 M solution of potassium acetate to pH 7.5 using KOH, then sterilise by filtration through a 0.22 μ membrane, and store frozen. Prepare a 10 mM potassium acetate, 10% glycerol solution using this stock and purest available water. Add the salts as solids, and adjust the pH (if necessary) to 6.4 using 0.1 N HCl. Do not adjust the pH upward with base. (The pH will drift for $1-2$ days before settling at $6.1-6.2$.) Sterilise the solution by filtration through a pre-rinsed 0.22 μ filter, and store at 4°C.

C. *Simple Transformation Buffers (SB)*

Compound	Amount/litre	Final concentration
SB		
KCl	7.4 g	100 mM
CaCl$_2$.2H$_2$O	7.5 g	50 mM
K-MES	20 ml of a 0.5 M stock (pH 6.3)	10 mM
(final pH 6.2)		
SB2		
KCl	7.4 g	100 mM
MnCl$_2$.4H$_2$O	8.9 g	45 mM
CaCl$_2$.2H$_2$O	1.5 g	10 mM
K-MES	20 ml of a 0.5 M stock (pH 6.3)	10 mM
(final pH 6.2)		

Table 10. *continued.*

D. *Simple Frozen Storage Buffer (FB)*

Compound	Amount/litre	Final concentration
KCl	7.4 g	100 mM
$CaCl_2.2H_2O$	7.5 g	50 mM
Redistilled glycerol	100 g	10% (w/v)
Potassium acetate (final pH 6.2)	10 ml of a 1 M stock (pH 7.5)	10 mM

Preparation of SB and FB: $CaCl_2$ is added to purest available water made 10 mM in the noted buffer, which has been prepared as described above. The buffers are not strictly required, but seem to improve reliability and reproducibility of the transformation. Adjust the pH as necessary to 6.2. 10 mM MOPS pH 6.5 may be used in place of the potassium acetate in FB.

E. $\chi 1776$ *Transformation Buffer (XFB)*

Compound	Amount/litre	Final concentration
RbCl	12 g	100 mM
$MnCl_2.4H_2O$	8.9 g	45 mM
Potassium acetate	35 ml of a 1 M stock (pH 7.5)	35 mM
$CaCl_2.2H_2O$	1.5 g	10 mM
$MgCl_2.6H_2O$	1.0 g	5 mM
LiCl	20 mg	0.5 mM
Sucrose (ultrapure) (final pH 5.80)	150 g	15% (w/v)

Preparation: Add all the compounds except for $MnCl_2$ and RbCl to the purest available water. Adjust the pH to 5.92. Then add $MnCl_2$ and RbCl as solids. The pH should now be $5.80 \pm .02$. It is best to avoid subsequent adjustment of the pH. Filter through a pre-rinsed 0.22 μ membrane and store the solution at $-20°C$.

F. *RF1*

Compound	Amount/litre	Final concentration
RbCl	12 g	100 mM
$MnCl_2.4H_2O$	9.9 g	50 mM
Potassium acetate	30 ml of a 1 M stock (pH 7.5)	30 mM
$CaCl_2.2H_2O$	1.5 g	10 mM
Glycerol (final pH 5.80)	150 g	15% (w/v)

Adjust the pH to 5.8 with 0.2 M acetic acid. Sterilise by filtration through a pre-rinsed 0.22 μ membrane.

G. *RF2*

Compound	Amount/litre	Final concentration
MOPS	20 ml of a 0.5 M stock (pH 6.8)	10 mM
RbCl	1.2 g	10 mM
$CaCl_2.2H_2O$	11 g	75 mM
Glycerol	150 g	15% (w/v)

Adjust pH to final pH 6.8 with NaOH (as necessary) and sterilise by filtration through a pre-rinsed 0.22 μ membrane.

Table 10. *continued.*

H. *DMSO and DTT Solution (DnD)*

Compound	Amount/10 ml final volume	Final concentration
DTT	1.53 g	1 M
DMSO	9 ml	90% (v/v)
Potassium acetate	100 μl of a 1 M stock (pH 7.5)	10 mM

I. *DTT Solution*

Compound	Amount/10 ml	Final concentration
DTT	3.3 g	2.2 M
Water	7 ml	
Potassium acetate	100 μl of a 1 M stock (pH 7.5)	100 mM

Preparation: Combine the components. Adjust the final volume to 10 ml using the appropriate solvent. Sterilise the aqueous DTT solution by filtration through a 0.22 μ filter on a 10 μl syringe. The DnD solution need not be sterilised by filtration provided that the potassium acetate stock solution is sterile. If filtration is necessary, then this should be carried out through Millipore Millex SR membrane unit, which is designed for use with organic solvents. Distribute aliquots to small tubes and store frozen. Individual aliquots may be thawed and refrozen several times. 10 mM MES pH 6.2 may be used in place of potassium acetate in these solutions.

[a]HACoCl₃ is hexamine cobalt (III) + trichloride.

χ1776 is extremely sensitive to soap, therefore it is essential to avoid all use of soap on glassware employed in transformation, and the additional use of disposable plastic tubes wherever possible is highly recommended. χ1776 dies rapidly on plates and in stabs, and is best stored as a frozen stock. The Xb medium (*Table 12*) has proved very good for growing χ1776, and it should be used in all growth conditions including large cultures for preparing plasmid DNA. Vigorous agitation is necessary for all liquid growth conditions whether to prepare either competent cells or plasmid DNA. Frozen preparations of competent χ1776 may be stored at −70°C. The freeze shock is simply extended to a storage phase. When needed, aliquots are thawed and carried through the remainder of the procedure. In this case, however, DNA is added after the freeze-thaw step.

5. MATERIALS

The materials employed in preparing competent *E. coli* are primarily common chemicals available from many sources. As a general rule reagent grade (or ultrapure) quality should be used. It is likely that virtually any supplier of chemicals will suffice for the cations. If transformation frequencies prove inadequate, then a programme of substitution should be instituted in order to identify the compound containing inhibitory impurities. Suppliers used by the author are listed in *Table 11* in order to allow other investigators to compare their commonly used chemicals to ones known to be of adequate purity.

The problems of reagent purity seem to apply primarily to the high efficiency protocols which employ organic solvents. The simple protocols are considerably less sensitive to impurities. The common sources of problems in the high efficiency method lie with quality of water and DMSO. These problems are dis-

Table 11. Examples of Suppliers[a].

Compound	Manufacturer	Stock number
KCl	Alpha, Danvers, Mass.	87626
	Fluka, Hauppauge, NY	60130
	Aldrich, Milwaukee, WI	20,409-9
$MnCl_2.4H_2O$	MCB, Cincinnati, OH	MX0185
	Aldrich	20,373-4
$CaCl_2.2H_2O$	Fisher	C79
$HACoCl_3$	Alpha	23144
	Aldrich	20,309-2
MES	Research Organics, Cleveland, OH	0113M
	Sigma, St. Louis, MO	M8250
DMSO	Fluka	41640
	Mallinkrodt	5507
	MCB, Omnisolve	MX1456
DTT	Calbiochem, La Jolla, CA	233155
Potassium acetate	Aldrich	24,038-9
Redistilled glycerol	BRL, Gaithersburg, MD	5514UA
	Alpha	13797

[a]Products from the indicated suppliers have repeatedly given good results in the hands of the author. This is not to imply that products of other manufacturers may not be acceptable or that the products of other manufacturers have been comprehensively tested.

cussed below. Other chemicals that have been problematic include MES, glycerol (for the frozen storage protocols), and dithiothreitol. Substitution experiments are the only way to identify impure reagents or other material problems. Additional problems encountered on occasion are discussed in Section 4.2.

Water purity can be a subtle problem, given that tapwater is acceptable for preparation of TFB in some locations, whereas in others even double distilled water is unsatisfactory. The most reliable water purification has proved to be the reverse osmosis system of Millipore Corp., which is effective in the removal of organic compounds. The water produced by this system can be seen to decline in transformation quality as the organic cartridges saturate, even while the conductivity remains very high. The fact that some stills produce unsatisfactory water suggests that organic compounds of similar volatility to water are the source of the observed inhibition of transformation. The effect is observed only in protocols employing solvents, which indicates that the solvent is interacting with organic contaminants and thereby producing undesirable effects. Treatment of water with activated carbon has proved to be of benefit in removing putative contaminants. Such treatment can render otherwise unsatisfactory water suitable for making excellent transformation solutions. This treatment is described in *Table 13*, and its trial use is recommended in cases where transformations are continually low.

The solvent DMSO undergoes an oxidation reaction, producing dimethyl sulphone and dimethyl sulphide. Dimethyl sulphide is a potent inhibitor of

Table 12. Media.

A. *SOB*

Bacto tryptone	2%
Bacto yeast extract	0.5%
NaCl	10 mM
KCl	2.5 mM
$MgCl_2$	10 mM
$MgSO_4$	10 mM

SOC

SOC medium is identical to SOB medium but contains 20 mM glucose in addition.

Preparation of SOB and SOC

Combine tryptone, yeast extract, NaCl and KCl in the purest water available. Autoclave for 30 − 40 min. Use within 2 − 3 weeks. Make a 2 M stock of Mg^{2+}, comprised of 1 M $MgCl_2$ plus 1 M $MgSO_4$. Sterilise by filtration through a 0.22 μ membrane. Prepare 2 M stock of glucose similarly, and store at − 20°C. Just prior to use, combine the medium with Mg^{2+} (and glucose for SOC) and sterilise by filtration through a 0.22 μ membrane. The final pH should be 6.8 − 7.0.

B. *X broth (Xb)*

Bacto tryptone	2.5%
Yeast extract	0.75%
$MgSO_4$	20 mM
Tris pH 7.5	50 mM
Diaminopimelic acid (DAP)	100 μg/ml
Thymidine	50 μg/ml

Preparation

Combine tryptone and yeast extract with the purest water available. Autoclave for 30 − 40 min. Prepare stocks of 2 M $MgSO_4$; 2 M Tris pH 7.5; and 100x [DAP (1 mg/ml) + thymidine (0.5 mg/ml)]. Sterilise each by filtration through a 0.22 μ membrane. Just prior to use, prepare Xb by combining the incomplete medium with the other reagents.

C. *LB medium*

Bacto tryptone	1%
Bacto yeast extract	0.5%
NaCl	1%

LM medium[a]

Bacto tryptone	1%
Bacto yeast extract	0.5%
NaCl	10 mM
$MgCl_2$	10 mM

D. *NZY medium*

NZ amine	1%
Bacto yeast extract	0.5%
NaCl	0.5%
$MgSO_4$	10 mM

E. *Psi medium*

Bacto tryptone	2%
Bacto yeast extract	0.5%
$MgSO_4$	20 mM
NaCl	10 mM
KCl	5 mM

Adjust the pH to 7.6 with KOH

[a]LM is generally used for selections on solid medium for the protocols described in this chapter. SOB solid medium is used primarily for developing colonies to be used in the preparation of competent cells.

Table 13. Water Purification.

1.	Take a 2−4 l Erlenmeyer flask and rinse thoroughly with water.
2.	Add water until the flask is half full. Any source of water − tap, dionized, distilled, Mill-Q, may be used.
3.	Add activated carbon, 100−325 mesh, at a ratio of 1% carbon to water (mass/mass).
4.	Place the flask on a magnetic stirring platform and stir for 1−2 days. The flask may be loosely capped with aluminium foil or a small beaker.
5.	Pass the water through filter paper to remove the carbon. The water can then be used for preparing transformation buffers, etc.

transformation (S.Benner and D. Hanahan, unpublished observations), and probably accounts for the observation that quality of DMSO is so important, and that storage of good quality DMSO eventually produces ineffective DMSO. The quality of DMSO stock is assessed by noting the enhancement of transformation frequency with its addition. This should give frequencies 10−20 times greater than in experiments in which it is not used. There are several alternative strategies for maintaining DMSO. The first is to find a good supplier of redistilled DMSO (usually spectrophotometric grade), purchase it in small quantities, and then aliquot an entire bottle into small polypropylene tubes. Each tube is completely filled with DMSO, and all are stored at −20 to −70°C. A tube is thawed and used on a given day and then discarded. When the combined DMSO and DTT solution is used, small aliquots should be taken, such that each need be thawed and refrozen less than five times. Storage at −70°C may retard the oxidation reaction.

An alternative or supplemental procedure is to bubble N_2 through a bottle of DMSO for 30− 60 min prior to its use. This seems to remove dimethyl sulphide and may restore unsuitable DMSO.

6. EVALUATION OF TRANSFORMATIONS

There are several standard controls which provide a reasonable basis for evaluating transformations:

6.1 State of Growth

Transformation frequencies peak when a culture of *E. coli* cells is in the mid-late phase of logarithmic growth. When the cells have been collected and while they are chilling on ice, determine the viable cell density. One way to do so is to dilute 10 μl of the culture into 1 ml of medium, 10 μl of that dilution (10^2) into 1 ml medium (now a 10^4 dilution), and take 10 μl of the 10^4 dilution and spread it on an agar plate of rich medium with no drugs. This gives a 10^6 dilution, such that 50 colonies correspond to a viable cell density of 5 x 10^7 cells/ml. For most strains of *E. coli*, the cell density should be in the range of 4−8 x 10^7 cells/ml. If the cell density varies significantly from this value, poor transformations may result. One generally assesses cell density by determining the apparent OD_{550} (which is actually light scattering) using a spectrophotometer. Each instrument should be calibrated to determine the relation between optical density and viable cell density, as considerable variation can occur among different brands of spectro-photometers, as well as with different strains of *E. coli*.

6.2 Transformation Frequency

The transformation frequency (or transformation efficiency, XFE) is determined using a DNA standard (such as pBR322) at levels well below saturation. This value gives a measure of how effectively individual competent cells take up and permit a plasmid to become established. For the high efficiency methods, 10 – 100 pg of pBR322 are used in a discrete transformation. Several fractions of the transformation are plated and the transformation frequency determined. For example, if 10 pg pBR322 are used, and 1% and 10% plated, then 10 colonies on the 1% plate or 100 colonies on the 10% plate correspond to an XFE of 1 x 10^8 colony forming units (c.f.u.)/μg pBR322. For the lower efficiency methods, 1 ng pBR322 should be used, again plating 1% and 10%. Here 10 colonies on the 1% plate or 100 colonies on the 10% plate correspond to a transformation efficiency of 1 x 10^6 c.f.u./μg.

6.3 Fraction of Competent Cells

The fraction of cells in a population that has been rendered competent for trans-formation is another significant parameter in evaluating transformation. It ranges from 0.01% to 10% of the viable cells. The fraction of competent cells indicates the total colony forming potential of that preparation, and is deter-mined with saturating amounts of plasmid DNA, typically using plasmid to cell ratios of 200:1. Thus, for a transformation of 10^8 cells, >2 x 10^{10} plasmid mol-ecules should be applied (100 ng pBR322). A good test is to use 200 ng pBR322 in a transformation, take 10 μl of that into 1 ml (10^2 dilution), 10 μl of the 10^2 di-lution into 1 ml medium, and then plate 1% and 10% of that on both rich plates (non-selective) and selective plates. The fraction of competent cells is then deter-mined by comparing the numbers of colonies formed under selective and non-selective conditions. With the standard high efficiency transformation, $>5\%$ of the cells should be competent.

6.4 Impact of Additives and Components

Those protocols employing additives can be tested by comparing the transform-ation efficiency (and the fraction of competent cells) obtained with and without their presence. DMSO should give a 10- to 20-fold enhancement, while DTT im-proves transformation by 2- to 4-fold. If cell densities are well off the optimal range, these values may be considerably less. Therefore, cell density should be determined when performing these controls.

The components of the transformation buffers can be best tested by substi-tution of the same reagents acquired from other manufacturers. It is worth trying standard tapwater that has been aerated to remove chlorine, instead of double distilled water or water purified by reverse osmosis. All sources of water can be extracted with activated carbon (*Table 13*). The components can also be separate-ly removed from the transformation bufer, and the results compared with those of Figure 1 of reference 15.

Controls can be divided into two classes: those used routinely and those used for troubleshooting. The determination of viable cell density and transformation efficiency should be undertaken with every transformation. These controls provide immediate feedback if frequencies are low. For example, if the cell density was 5×10^8, and the XFE 2×10^7 using the high efficiency protocol, one can identify the reduced efficiency as excessive cell density. The remainder of the controls are used for troubleshooting when transformations are consistently inadequate. In the experience of the author, a series of substitution experiments will provide reasonably efficient identification and solution to the problems.

7. GENERAL PRACTICE

There are several general techniques of relevance to plasmid transformation and these are noted briefly below:

7.1 Plating

Cells carried through the transformation protocols are somewhat fragile and are best treated gently during their treatment immediately thereafter. One reliable plating method is to spot 150 μl of medium on the plate, then pipette the desired amount of cells into the puddle. (If the cell volume being plated is >200 μl, then this is not necessary.) The cells are spread quickly but gently using a bent pasteur pipette. The cells should be spread in liquid, and not on a dried plate. The author uses a plating wheel. The plate is spun slowly and the L-shaped pasteur pipette rested lightly on the plate, thus drawing the pool of liquid evenly over the surface of the plate.

If the entire transformation mixture or the mixtures from several tubes are to be spread onto one plate, the cells are first pelleted by centrifugation for 5 min at $600-800$ *g*, the liquid poured off, and the cells suspended in $100-300$ μl medium by gentle pipetting up and down.

7.2 Plating in Top Agar

One or more entire transformation mixtures can be easily established on a single plate by using top agar. A solution of SOB medium (*Table 12*) plus 1% agar or agarose is melted and kept at $42-45°C$. Two ml of top agar are added to the cell suspension, and the solution is mixed and poured onto a plate. After allowing the agar to solidify, the plates are incubated at 37°C to establish colonies. If the agar is solidifying before being dispersed evenly on the plate, warm both plate and transformation mixture to 37°C just before plating.

7.3 Heat Shock

The heat shock step is of general applicability to all the techniques for transforming *E. coli*. The optimal times of heat shock are in part a function of the surface to volume ratio of the cell suspension and the thermal conductivity of the tubes used. If scaled-up transformations are being performed in large tubes, or if materials other than polypropylene are being used, it is advisable to calibrate the heat shock time at 42°C to determine the best value.

7.4 Storage of Transformed Cells

Transformed cells may be stored overnight at 4°C following the 1 h incubation at 37°C. The titer of transformed cells declines to about 90% after 12 h, provided that SOC medium (*Table 12*) is used for the incubation. The presence of 20 mM glucose improves the viability of transformed cells. For longer storage, the cells should be washed to remove the transformation bufer. Add 2 ml of SOC medium and pellet the cells at 800 *g*. Decant the supernatant and gently resuspend the cells in SOC medium. The titer will remain at >90% for several days of storage at 4°C.

For the long term storage of transformed cells, dilute the transformation mixture 1:1 with a solution of 40% glycerol, 60% SOB medium (to give a final concentration of 20% glycerol), transfer to screw cap tubes, chill on ice, flash freeze in liquid nitrogen, and place the frozen culture at −70°C. The titer after thawing will be >90%.

8. SPECIAL APPLICATIONS

8.1 cDNA Cloning

The conversion of mRNA into double stranded cDNA and its insertion into an appropriate plasmid vector is in sum a difficult and inefficient process. It is therefore not unusual to have a small amount of material from which a representative library must be created. Thus, the total colony forming potential of a cDNA preparation must be maximized. This requires high transformation frequencies, high proportions of competent cells in the transformation cocktails, and transformation under conditions that optimize both the total number of colonies and the density of colonies formed on each plate. These last parameters must often be balanced, as working at very low DNA levels will maximize the total number of colonies formed, but will as a consequence require more discrete transformations, of which only about two can be plated on a 100 mm agar plate.

An effective way to optimize colony forming potential and colony density is to determine the saturation level for a given preparation of cDNA plus vector. Different preparations will give different saturation points, even though they contain equal masses of total DNA. An example of this is shown in *Figure 1*, where several similar cDNA preparations are compared. Each has been used in transformations in amounts ranging from 0.25 ng to 100 ng total DNA. For each preparation a different amount of DNA can be taken as a good compromise between total colony forming potential (size of the library) and colony density in each transformation. It is of particular note that all of these preparations are saturated at 30 ng total DNA. It is common to view 50−100 ng of cDNA plus vector as a small amount and thus apply that quantity to a transformation. As can be seen, about 10-fold more colonies would be produced by performing 10 discrete transformations with 5 ng DNA. Therefore, it is advisable to determine a saturation curve for each cDNA preparation prior to establishing it in *E. coli*.

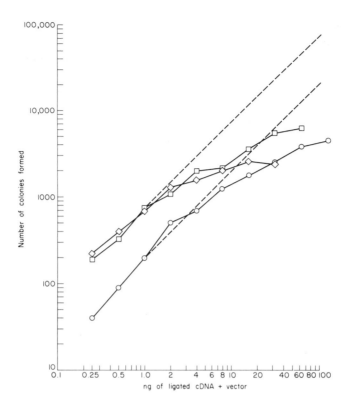

Figure 1. Linearity of transformation with ligated cDNAs. Three different preparations of cDNAs were used in a series of transformations in which varying levels of DNA were combined with a fixed volume of competent cells. Two of the cDNA preparations (□ and ○) were derived from human placental mRNA and were ligated to an *Eco*RI-*Sal*I pBR322 vector at a 25:1 vector to cDNA mass ratio (approximately to a 5-fold molar excess of vector). The third (◊) was from embryonic chick smooth muscle mRNA and was ligated to pUC9 at a 20:1 mass ratio. In all cases the noted amount of ligated DNA was used in a standard transformation, in which supercoiled pBR322 gave 5 x 10^8 transformants per μg DNA. The dotted lines are extrapolations of the expectation that a doubling of the amount of DNA should double the number of colonies. Both scales on the graph are logarithmic.

8.2 Plasmid Rescue

Retrieving plasmids, or 'plasmid rescue', from integrated states in high mol. wt. cellular DNA is achieved by (i) digestion with a restriction endonuclease that does not cleave within the drug resistance gene(s) or replication origin of the plasmid, (ii) ligation at low DNA concentration to cyclize the digestion products, and (iii) transformation of *E. coli*. In this process one seeks to select the rare cellular fragment carrying an integrated plasmid by its ability to transform *E. coli*.

This technique has been applied to retrieve integrated plasmids and flanking DNA from yeast (20), cultured mammalian cells (21,22), and, more recently, from transgenic mice (D.H., unpublished observations). Plasmid rescue from mammalian cells has proved very inefficient (22), in that the transformation frequencies obtained are inconsistent with the mass of plasmid DNA present. This necessitates high transformation efficiency in order to counterbalance the inefficiency of rescue.

In plasmid rescue, most of the DNA is nontransforming and competes with the plasmid DNA. Competition with nontransforming DNA becomes appreciable at a DNA concentration of about 100 ng. Although in principle one should use total DNA levels where no competition occurs, a good compromise level which minimizes the number of separate transformations required for rescue has been found to be in the range of 200 − 500 ng total DNA per discrete transformation.

When cellular DNA is prepared for rescue, the ligation should be assessed as follows. Aliquots of the restricted DNA and the restricted and ligated DNA are electrophoresed side by side through a 0.75% agarose gel containing no ethidium bromide. The gel is stained with ethidium bromide, which is applied after electrophoresis. The ligated DNA should show both an appreciable shift to higher mol. wt., and DNA migrating well above the limiting mobility for linear DNA. These molecules are relaxed closed circles, and are particularly useful for assessing the degree of cyclization of the ligated DNA. As a general rule, ligations for plasmid rescue should be performed at DNA concentrations of $10 - 20 \mu g/ml$, and the DNA then precipitated and resuspended to a concentration of $25 - 100 \mu g/ml$. If no closed circular DNA can be detected using this assay, it is probable that the DNA has not been adequately cyclized and is thus not likely to be useful for efficient plasmid rescue. In this case different preparations of the restriction enzymes or DNA ligase should be tried.

8.3 Storage and Amplification of Plasmid Libraries

Populations of recombinant plasmids can be introduced into *E. coli* using the methods described in this chapter and then maintained as distributions of colonies on a porous support of nitrocellulose. Such distributions can be replicated accurately, and multiple copies produced to be used in nucleic acid hybridization or antigenic screening assays, as well as for extended storage (23,24).

Maintaining and copying colonies on nitrocellulose filters has the advantage that it maintains each clone in one location, and thus prevents bias in the representation of each clone resulting from differential growth rates. In contrast, growth of a population of recombinants in liquid culture can allow certain classes of clones to become significantly over- or under-represented in the amplified population. Growth on filters seems to retard excessive growth of fast-growing clones, presumably due to the physical environment of a colony, which is likely to be less favourable to rapid and continued growth than that of an isolated cell in a liquid environment. A major disadvantage of maintaining colonies on filters is that replication of filters is laborious, and this precludes convenient production of many copies of a plasmid library, and probably inhibits the distribution of valuable clone banks to other investigators.

A reasonable compromise between the benefits of amplification on filters and the ease of distribution of liquid cultures is to do both. Clones are amplified as colonies on filters, which prevents significant overgrowth of favourable (often non-recombinant) clones. The colonies on the filters are then dispersed and mixed in liquid, and aliquots frozen at $-70°C$ or in liquid nitrogen. One filter of 20 000

Table 14. Amplification and Storage of Plasmid Libraries for Distribution.

A. *Preparation*

1. Take the plasmid library, which is being maintained as a collection of colonies on nitro-cellulose filters, and prepare a fresh set of replicas as described in references 23 and 24. Incubate the replicas at $30 - 37°C$ until the colonies are $0.5 - 1$ mm in diameter assuming a density of $5000 - 20\ 000$ colonies per filter. It is recommended that the plating, replication and subsequent steps be performed in a sterile hood, so as to minimize contamination with other organisms.

2. Transfer each filter with the colonies uppermost into an empty Petri dish. Add 10 ml of SOB medium containing 20% glycerol. Disperse the colonies by rubbing the filter with a rubber policeman.

3. Collect the cell suspension by pipetting up and down and then transfer it to a sterile tube and place on ice. Vortex to complete the dispersal. The cell density should be about 10^9 per ml, and at this point each filter set of colonies has been dispersed in a separate tube. These sets of cells can either be combined at this stage to produce a complete mixture of the library, or maintained separately as N sets derived from the N master filters of colonies that carried the library.

4. Distribute $100 - 500\ \mu l$ volumes to polypropylene tubes. Flash freeze in liquid nitrogen or a dry ice/alcohol bath, and place in liquid nitrogen or at $-70°C$.

B. *Retrieval of frozen aliquots of an amplified library*

1. Remove an aliquot (or a set of aliquots) that represent the library. Thaw in air until just liquid and then place on ice.

2. Dilute the aliquot(s) 10x into SOC medium, producing a set of primary aliquots.

3. Prepare a set of serial dilutions of the primary aliquot(s), so as to determine colony forming potential. Since the original cell density was about $10^9/ml$, the primary aliquots should now have a density of about 10^8 cells/ml, assuming no cell death. However, the titer of viable cells will decline slowly upon storage, and therefore the titer will be less than that when collected.

4. Plate 10 μl aliquots of the 10^2, 10^3, 10^4 and 10^5 serial dilutions from the primary aliquots onto plates carrying the appropriate selective drug. Incubate overnight to establish colonies. Store the primary aliquot(s) on ice, and discard the serial dilutions.

5. The titer of the primary aliquots can be determined from the platings of the dilutions. The titer of the primary aliquot(s) will remain $>90\%$ of its original value during 24 h of storage on ice. Prepare a fresh dilution from the primary aliquot such that there are about 10^5 colony forming units per ml. Thus, $50 - 250\ \mu l$ can be applied to each plate, to produce a colony density of $5000 - 25\ 000$. Plate on nitrocellulose filters, incubate to establish colonies, and replicate the distributions as described in references 23 and 24. Since the original distribution of colonies has been dispersed and thus randomized, more filters should be prepared than the original library required, so as to assure representation of all clones. A 100 μl aliquot derived from a filter carrying 10^4 colonies will initially have 10^7 cells in it, or about 1000 copies of the distribution. Even very long term storage should allow that aliquot to restore the clones carried in the original distribution.

---

1 mm diameter colonies contains about 10^{10} cells. If these are dispersed in 10 ml and 100 aliquots frozen down, each aliquot will initially contain 10^8 cells (or ~ 5000 copies of each clone). Thus even as the titre of usable cells declines during storage or shipment, sufficient colony forming potential will remain to produce many copies of the original distribution. This is analogous to the plate lysate method for amplifying bacteriophage libraries (25).

This approach will of course randomize the original colony distribution, and thus not every clone will be represented on a replating of the same number of colonies from the amplified stock. The statistics of this can be modeled by the Poisson distribution, given the assumption of no bias between individual clones (an often inaccurate but necessary approximation). Thus, if a library maintained on N filters is mixed and subsequently replated in order to re-establish it, some 1.5 − 2 N filters should be produced to assure representation of all clones in the original population. For a discussion of this, see Clarke and Carbon (26). The price of screening additional filters seems likely to prove reasonable for the safe maintenance and distribution of rare and valuable cDNA libraries. *Table 14* presents a protocol which illustrates one way of preparing and retrieving cDNA banks in this manner.

9. REFERENCES

1. Cohen,S.N., Chang,A.C.Y. and Hsu,L. (1972) *Proc. Natl. Acad. Sci. USA,* **69**, 2110.
2. Mandel,M. and Higa,A. (1970) *J. Mol. Biol.,* **53**, 159.
3. Lederberg,E.M. and Cohen,S.N. (1974) *J. Bacteriol.,* **119**, 1072.
4. Taketo,A. (1972) *J. Biochem.,* **72**, 973.
5. Taketo,A. (1974) *J. Biochem.,* **75**, 59.
6. Taketo,A. and Kuno,S. (1974) *J. Biochem.,* **75**, 895.
7. Kretschmer,P.J., Chang,A.C.Y. and Cohen,S.N. (1975) *J. Bacteriol.,* **124**, 225.
8. Enea,V., Vovis,G.F. and Zinder,N.D. (1975) *J. Mol. Biol.,* **96**, 495.
9. Kushner,S.R. (1978) in *Genetic Engineering,* Boyer,H.B. and Nicosia,S. (eds.), Elsevier, Amsterdam, p. 17.
10. Norgard,M.V., Keem,K. and Monahan,J.J. (1978) *Gene,* **3**, 279.
11. Dagert,M. and Ehrlich,S.D. (1979) *Gene,* **6**, 23.
12. Morrison,D.A. (1979) in *Methods in Enzymology,* Vol. **68**, Wu,R. (ed.), Academic Press, Inc., NY, p. 326.
13. Jones,J.M., Primrose,S.B., Robinson,A. and Ellwood,D.C. (1981) *J. Bacteriol.,* **146**, 841.
14. Bergmans,H.E.N., van Die,I.M. and Hoekstra,W.P.M. (1981) *J. Bacteriol.,* **146**, 564.
15. Hanahan,D. (1983) *J. Mol. Biol.,* **166**,
16. Appleyard,R.K. (1954) *Genetics,* **39**, 440.
17. Casadaban,M. and Cohen,S. (1980) *J. Mol. Biol.,* **138**, 179.
18. Meselson,M. and Yuan,R. (1968) *Nature,* **217**, 1110.
19. Curtiss,R., III, Inoue,M., Pereira,D., Hsu,J.C., Alexander,L. and Rock,L. (1977) in *Molecular Cloning of Recombinant DNA,* Scott,W.A. and Werna,R. (eds.), Academic Press, NY, p. 99.
20. Hicks,J.B., Hinnen,A. and Fink,G.R. (1979) *Cold Spring Harbor Symp. Quant. Biol.,* **43**, 1305.
21. Peucho,M., Hanahan,D., Lipsich,L. and Wigler,M. (1980) *Nature,* **285**, 207.
22. Hanahan,D., Lane,D., Lipsich,L., Wigler,M. and Botchan,M. (1980) *Cell,* **21**, 127.
23. Hanahan,D. and Meselson,M. (1980) *Gene,* **10**, 63,
24. Hanahan,D. and Meselson,M. (1983) in *Methods in Enzymology,* Vol. **100**, Wu *et al.,* (eds.), Academic Press, NY, p. 333.
25. Maniatis,T., Hardison,R., Lacy,E., Lauer,J., O'Connel,C., Quon,D., Sim,G.K. and Efstradiadis,A. (1978) *Cell,* **15**, 687,
26. Clarke,L. and Carbon,J. (1976) *Cell,* **9**, 91.

CHAPTER 7

The Use of Genetic Markers for the Selection and Allelic Exchange of in Vitro Induced Mutations that do not have a Phenotype in E. Coli

GIANNI CESARENI, CINZA TRABONI, GENNARO CILIBERTO, LUCIANA DENTE and RICCARDO CORTESE

In this chapter we will describe two sets of techniques which facilitate the isolation of mutations in preselected DNA fragments and their insertion in the correct genetic context. The techniques can be used to select mutants that do not have a phenotype in *E. coli*. The mutations are tagged by linking them to genetic markers that can be easily identified in *E. coli*.

1. A GENERAL METHOD FOR INDUCTION AND SELECTION OF BASE-PAIR SUBSTITUTIONS

In a project aiming to identify DNA sequences which are important for a specific function, it is often necessary or useful first to define the DNA region of interest. It is important to use rapid and precise methods for exploratory mutagenesis to achieve this. Guided by the results so obtained one can then use very specific methods of *in vitro* mutagenesis such as those based on synthetic deoxyoligonucleotides (1).

We have recently developed a simple procedure which allows the rapid isolation and identification of a large number of base-pair substitution mutants in any preselected region of DNA. The method exploits the property of reverse transcriptase to incorporate non-complementary bases during DNA synthesis *in vitro*. It is designed to associate a distinguishable phenotype with bacterial clones carrying the desired mutations (2,3). The method is best applied when the target DNA sequences is cloned in a single-stranded DNA vector. As described in Chapter 5 of this book, it is possible to convert any plasmid into a single-stranded DNA vector, using simple manipulations. These protocols for mutagenesis are therefore universally applicable (4).

1.1 The Mutagenesis Step

A segment of DNA is annealed to a single-stranded DNA template in order to prime DNA synthesis *in vitro*. During elongation in the presence of an incomplete set of deoxynucleotide triphosphate, misincorporation of an incorrect nucleotide can occur if the correct nucleotide is absent (5). This 'mistake' cannot be corrected by polymerases lacking a 3' exonuclease activity, such as reverse transcriptase. Once the misincorporation has occurred, the primer, which now carries a mismatched nucleotide at its 3' end, can be elongated in the presence of the correct nucleotides (3,5). In this way, it is possible to generate any base-pair change in the position adjacent to the 3' of the primer and this will usually be the only base-pair change introduced. If however, the primer-template complex is incubated in the presence of reverse transcriptase and an incomplete set of deoxynucleotides, multiple base-pair substitutions can be generated at a template position complementary to the omitted nucleotides.

1.2 **Mutants Selection Protocol**

Borrias *et al.* (6) have shown that it is possible to select wild-type ϕX174 phage particles following transformation of *E. coli* with single-stranded ϕX174 DNA carrying an amber mutation annealed with a short ϕX174 DNA segment containing the corresponding wild-type sequence. Inside the cell this wild-type primer is elongated so as to complete the complementary strand. The resulting double-stranded DNA is replicated, yielding a progeny of wild-type and mutant viruses. We have exploited this observation by associating a selectable marker with the primer used in the *in vitro* mutagenesis reaction. This selectable marker will therefore be associated with the DNA strand synthesised *in vitro* and so linked to any mistake made during elongation of the DNA chain (3,4).

It is possible to use a variety of selectable markers. We will present a detailed protocol that we have used for extensive *in vitro* mutagenesis of a eukaryotic tRNA gene. The phenotype that we used in this series of experiments, in order to distinguish template derived from primer-derived M13 phages, relies on the ability of a functional α-peptide of the enzyme β-galactosidase to complement the α-fragment and to metabolise 5-chromo-4-chloro-3-indolyl-β-D-galactoside (X-gal) thus forming a blue precipitate. An important property of the α-peptide is that its amino terminus is not important for its function and can be substituted by any amino acid sequence. This property has allowed us to develop a protocol to introduce base-pair substitutions in any short DNA segment and to select clones which carry the mutations. The procedure is the following:

(i)   The wild-type DNA segment to be mutagenised is cloned in an M13mp phage vector, within the coding sequence of the α-peptide of β-galactosidase so as to alter the translational reading frame. A phage carrying such an insertion will be unable to produce active β-galactosidase and so will form white plaques on X-gal indicator plates (7).

(ii)   The wild-type single-stranded DNA is annealed with a short DNA segment carrying insertions or deletions of one or two bases that restore the correct translational reading frame and whose 3' end is complementary to sequences located in the proximity of the region where base-pair substitutions are to be introduced. It is important to realise that the insertions or deletions which restore the reading frame can be outside the gene to be mutagenised, and conveniently placed in the adjacent vector sequences.

(iii)   The primer is elongated in the presence of reverse transcriptase and a subset
        of deoxynucleotide triphosphates to force the incorporation of a non-
        complementary nucleotide at the desired position. A full complement of
        deoxynucleotide triphosphates is then added to continue elongation. The
        annealed complex is then used to transform *E. coli*. The progeny of the
        template strand and the primed strand yield white and blue plaques respec-
        tively. Blue plaques should contain the desired mutations. Since the primer
        is often obtained by restriction enzyme digest, in many cases it does not
        have its 3'-OH end in the ideal position to induce the desired mutation.
        In this situation it is necessary to modify the primer by lengthening or shorten-
        ing it. This can usually be accomplished by exploiting the polymerase or
        exonuclease activity of *E. coli* DNA polymerase Klenow fragment, and
        an opportune choice of a subset of deoxynucleotides (2).

An example of the use of this protocol is depicted in *Figure 1*. The template
(prepared as described in *Table 1*) was a segment of DNA coding for a eukaryotic
tRNA^{Pro} which was cloned between the *Eco*RI and *Bam*HI sites of M13 mp701
(2). This resulted in a recombinant, called R78B, which yielded white plaques
because the fragment inserted into the coding region of the β-galactosidase gene
caused a -1 frameshift in the reading frame. The primer was obtained from
another plasmid, R78Bo. This plasmid was constructed by filling in the cohesive
*Bam*HI terminus of R78B. This addition of four bases restores the correct reading
frame for β-gal. The plaques formed by R78Bo are therefore blue (2). The protocol for the preparation of the primer is given in *Table 2*. A *Pst*I-*Hin*fI fragment
was purified from R78Bo and annealed to single-stranded R78B. The difference
in the sequence between primer and template (which results in the frameshift)
lies outside of the tRNA^{Pro} coding segment. We wished to substitute the G residue
in position 31 with a C residue. The presence of G residues in positions 27 and
29, between the 3' end of the primer (position 23) and the preselected site made
this substitution difficult. We therefore incubated the primer-template complex
with *E. coli* DNA polymerase Klenow fragment and a mixture of dATP, dTTP
and dCTP to add six bases to the primer so as to copy the G27 and G29 residues.
The Klenow fragment was then inactivated by heating at 75°C for 20 min and
the unincorporated nucleotides were removed by gel filtration. At this point the
3'OH of the primer was complementary to nucleotide G29 and in order to obtain
the G31 to C31 mutation it was sufficient to add reverse transcriptase and dGTP
to direct the correct incorporation of a G residue at position 30 and the misincorporation of a G residue at position 31. After 5 min, the remaining complement
of deoxynucleotides was added and DNA synthesis was continued for 30 min.
A detailed protocol for the *in vitro* synthesis is given in *Table 3*. The products
of these reactions were introduced into *E. coli* by the transformation protocol
in *Table 4*. DNA was made from two phages that gave blue plaques. Analysis
by DNA sequencing revealed the expected mutations.

1.3 Efficiency of Mutagenesis

In general the efficiency of these various reactions is such that, following transfor-

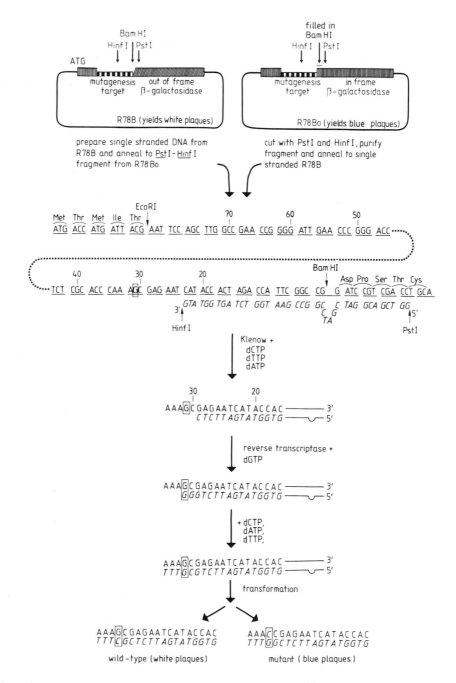

Figure 1. *In vitro* selection of base-pair substitutions induced *in vitro* by reverse transcriptase. The continuous line represents M13 vector sequences. Lined boxes indicate β-galactosidase coding sequences. The oblique lines in R78B represent the coding sequence shifted out of frame by the insertion of the mutagenesis target. Plasmid R78Bo is a derivative of R78B, where four bases have been added by filling in the *Bam*HI site; as a consequence the correct β-galactosidase frame has been restored (straight line box). Other details are described in the text.

Table 1. Preparation of Single-stranded DNA from M13-infected Bacteria.

The genome of M13 phages consists of a circular molecule of single-stranded DNA which is contained in a rod shaped virion. The phage is secreted from the host into the medium where it can reach a titer of approximately 10^{12} virions/ml. The single-stranded DNA can be easily purified from small volumes of supernatant of a bacterial culture infected by the phage. DNA is prepared in this way is suitable for DNA sequencing or for site-specific mutagenesis.

1. Pick a fresh plaque into 1.5 ml of L-broth using a sterile Pasteur pipette. Use a large glass tube and culture with aeration for 6 h at 37°C.
2. Pour the culture into an Eppendorf tube. Centrifuge for 5 min.
3. Pour off the supernatant into another Eppendorf tube, making no effort to transfer it completely. If you store the supernatant for more than a day, centrifuge it again before proceeding.
4. Add 200 μl of 2.5 M NaCl, 20% polyethylene glycol 6000. Leave at room temperature for 15 min.
5. Centrifuge for 3 min. Remove the supernatant with a Pasteur pipette. Centrifuge again briefly. Remove the remaining traces of supernatant with a drawn-out Pasteur pipette.
6. Add 100 μl of 10 mM Tris-HCl pH 8.0, 1 mM EDTA and resuspend the pellet. Add 100 μl of phenol saturated with 100 mM Tris-HCl pH 8.0 and 10 mM EDTA and extract the protein by gentle mixing.
7. Centrifuge for 1 min. Withdraw about 80 μl of the aqueous phase and add to a new Eppendorf tube.
8. Add 20 μl of 5 M sodium perchlorate and 50 μl of isopropanol.
9. Spin for 15 min in a microcentrifuge. Do not worry if you do not see any pellet.
10. Remove the supernatant with a drawn out pipette and dissolve the pellet in 100 μl of 0.3 M sodium acetate.
11. Add 300 μl of ice-cold ethanol. Leave for 5 min in a bath of dry ice and ethanol.
12. Spin for 15 min. Dissolve the pellet in 30 μl of 10 mM Tris-HCl pH 8.0, 0.1 mM EDTA.

Table 2. Preparation of Double-stranded DNA.

The protocol in this table uses the example in *Figure 1* in which a 39 base-pair fragment was purified from R78Bo DNA.

1. Digest about 10 μg of double-stranded R78Bo DNA with *Pst*I and *Hinf*I.
2. Load the digest onto a 15% polyacrylamide gel (1 mm thick, 20 cm wide and 20 cm long).
3. Electrophorese the gel at 150 V, 50 mA for 3 h.
4. Stain the gel for 15 min in 5 μg/ml ethidium bromide.
5. Cut out the 39 bp band and elute this for 5 h at 37°C with 5 ml of extraction buffer[a].
6. Dilute the sample 4-fold with water and apply to a 0.2 ml column of DEAE cellulose (DE 52).
7. Elute the fragment with 0.5 ml of 1.5 M ammonium acetate (pH 10.0) directly into a siliconised polyallomer tube for an SW60 rotor.

Note: It is important *not* to add carrier tRNA.

8. Add 1 ml of ice cold ethanol.
9. Centrifuge in the SW60 rotor for 15 min at 40 000 r.p.m.
10. Recover and dry the pellet in a vacuum centrifuge. Resuspend it in 10 μl of water.

[a]Extraction buffer is 0.5 M ammonium acetate, 15 mM magnesium acetate, 0.5 mM EDTA, 10% (w/v) SDS.

mation, only a few percent of the clones contain mutations, and therefore more work must be done to identify interesting mutants. The method presented in this chapter is designed so that any mutation is linked to a detectable marker. In this way, following transformation, the mutant clones are easily identified by their linked phenotype.

The frequency of mutants depends solely on the efficiency of the mutagenic

Table 3. The Mutagenesis Reaction.

A. *The Annealing Reaction*
1. Set up the following annealing mixture:
   4 μl Primer (fragment prepared as in *Table 2*)
   4 μl single-stranded circular DNA (150 μg/ml; R78B DNA in this example)
   1 μl *Hin* buffer (10x concentrated[a])
   1 μl water
2. Seal this in a siliconised capillary tube.
3. Boil for 3 min in a water-bath. Cool slowly at room temperature.

B. *Elongation of the Primer*
4. Incubate 5 μl of the annealed reaction mixture for 15 min at room temperature with 0.5 units of DNA polymerase (Klenow fragment) in Hin buffer, containing dCTP, dTTP and dATP at a concentration of 50 mM, in a final volume of 20 μl.
5. Heat at 75°C for 20 min.
6. Prepare a 0.8 ml column of Sephadex G50 in a 1 ml syringe connected to a needle. Insert the needle into the cap of a 1.5 ml Eppendorf tube. Centrifuge for 5 min at 1500 r.p.m. to remove excess water.
7. Apply the 55 μl reaction mixture to the column and connect the syringe to a new Eppendorf tube.
8. Spin for 5 min at 1500 r.p.m.

C. *Mispairing Reaction*
9. Set up the following mixture:
   5 μl of the primer-template annealed complex
   5 μl 5 mM dGTP
   5 μl Reverse transcriptase buffer (10x concentrated[b])
   35 μl water
   0.4 μl Reverse transcriptase (12 u/μl)
10. Incubate for 1 h at 37°C.
11. Add 5 μl of a 5 mM mixture of dATP, dCTP, dGTP, dTTP. Incubate for 1 h at 37°C.
12. Remove unincorporated nucleotides by gel filtration as described in steps 6−9.

[a]*Hin* buffer (10x concentrated) is 0.5 M NaCl, 66 mM $MgCl_2$, 66 mM Tris-HCl pH 8.0, 66 mM β-mercaptoethanol.
[b]Reverse transcriptase buffer (10x concentrated) is 0.4 M Tris-HCl pH 8.0, 0.4 M $MgCl_2$, 10 mM dithiothreitol.

Table 4. Transformation.

1. Add 20 μl of reverse transcriptase reaction mix to 200 μl of 71/18[a] or JM105[b] competent cells[c]. Incubate 1 h in ice.
2. Add 50 μl of freshly-grown saturated culture of 71/18 or JM105 cells, 25 μl of isopropyl β-D-thiogalactopyranoside (IPTG) (20 mg/ml) 25 μl of 5-brom-4-chlor-3-indolyl-β-D-galactoside (BCIG) (20 mg/ml).
3. Mix the transformation reaction with 3 ml of top[c] layer agar at 42°C and plate on BBL[c] plates.
5. Incubate overnight at 37°C.

[a]The genotype of *E. coli* 71/18 is Δ[lac-*pro*]F'[*lac*I[a]*lac*ZΔM15 *pro*AB]supE.
[b]The genotype of *E. coli* JM105 is Δ[*lac-pro*] *thi strp*A *end*A *sbc*B15 F' [*lac*I[q]lacZΔM15 *pro*AB$^+$].
[c]Competent cells are prepared from a bacterial suspension growing exponentially in 40 ml of L-broth. At O.D.$_{600}$ 0.2−0.4 cells are harvested by centrifuging at 3000 r.p.m. in a bench centrifuge. The pellet is gently resuspended in 20 ml of ice-cold 0.1 M $CaCl_2$ and incubated for 20 min at 0°C. After a second centrifugation and resuspension in 0.4 ml of 0.1 M $CaCl_2$ the competent cells may be used directly in the transformation experiment or kept in ice for a few days. See Chapter 6 for a full discussion of the preparation of competent cells.
[d]BBL plates are poured from a solution containing Bactotryptone 10 g, NaCl 5 g, Bacto-agar 10 g per liter. The concentration of agar in the top layer is 0.7% rather than 1%.

event and on the mismatch repair activity of *E. coli*. We have minimised the effect of the latter by using an under-methylated template DNA (2,15). According to our results, the efficiency of introduction of the incorrect nucleotide by reverse transcriptase under the experimental conditions described cannot be far from 100% because about 85% of all plaques selected on the basis of the primer-associated genetic marker contained mutations.

1.4 A More General Selective Marker

The frame-dependent phenotype described above has been very useful for the mutagenesis of relatively short DNA sequences. In the case of longer DNA fragments, however, this method is often not applicable because of the occurrence of nonsense codons in the sequence to be mutagenised. A more general approach is based on a movable selective marker for example, the 203 bp segment coding for *E. coli* tRNA^{tyr} *sup*F (8). This can be used as follows. The gene to be mutagenised is cloned in a single-stranded cloning vector. Double-stranded plasmid DNA is then treated with any chosen restriction enzyme so as to obtain a single cut per molecule and this mixture is ligated to purified tRNA^{tyr} *sup*F DNA. After transformation and selection for a recombinant carrying a suppressor gene, a family of recombinant plasmids is obtained which contains the tRNA^{tyr} insert at various positions along the sequence of the DNA to be mutagenised (3). Each of these plasmids can be used as a *template* for the *in vitro* mutagenesis reaction. The primer is a DNA segment obtained from the unmodified gene and chosen so that in each case it spans the regions flanking the tRNA^{tyr} insert. Base-pair substitutions are introduced using reverse transcriptase as described above. In this case mutants can be selected for the absence of tRNA^{tyr} *sup*F activity in a bacterial host containing appropriate selective markers carrying amber mutations. The selected recombinant will contain the desired mutation, but not the tRNA^{tyr} *sup*F DNA.

1.5 Alternative Methods for Exploratory Mutagenesis

A variety of alternative methods to construct base-pair substitution mutants have been described.

The single-stranded specificity of the mutagenic action of sodium bisulphite has been extensively used (9−12). This method has, however, two strong limitations: only base-transitions can be obtained and it is difficult to control the reaction so as to obtain single point mutations.

Mutagenesis based on the incorporation during *in vitro* DNA synthesis of base analogues with a potential for mispairing has also been used (13). In this case, as for the bisulphite mutagenesis, the limitation is that only base-transitions can be obtained.

Shortle *et al.* have used a method based on the forced incorporation of a wrong nucleotide during DNA synthesis *in vitro* with *E. coli* polymerase and exploiting the property that incorporated α-thionucleotides cannot be removed by the 3'-exonuclease activity of the Klenow enzyme (14). Their method is conceptually similar to the forced misincorporation using reverse transcriptase, as described in this paper and originally used by Zakhour and Loeb (5). The latter, however, has the advantage of using commonly available reagents.

2. A GENERAL METHOD TO EXCHANGE ALLELIC SEQUENCES BETWEEN TWO REPLICONS

The ultimate aim of a mutagenesis experiment is the study of the phenotype of a cell carrying the altered DNA sequence inserted in the correct genetic context. However, most of the mutagenesis methods, including that described in Section 1, are better applicable to relatively short DNA fragments cloned in appropriate vectors. The task of reconstructing the genomic sequence, essential for studying the effect of the mutation on the function of the gene of interest, is sometimes facilitated by the presence of suitable restriction enzyme sites. If these sites do not occur too frequently in the region of interest, and if they are located in convenient positions, they may allow reassembly of the genomic region. More generally a double recombination event, *in vivo*, allows the exchange of allelic sequences between two replicons. However, double recombination, is very rare and, unless a selection method is available, this procedure as such is not applicable. Fortunately a series of expedients have been used which overcome the problem of low efficiency by separating the double recombination event into an integration and an excision step (*Figure 2*). If both these recombination steps can be sequentially selected for, the efficiency of exchange of allelic sequences between the two replicons would depend only on the position of the mutation with respect to the two joints A and B between the inserted sequence and the vector. Most of these methods require that one of the two replicons carries a selectable marker and a replication origin which can be conditionally blocked (*Figure 2*). Selection for integration is achieved by selecting for this marker under conditions in which

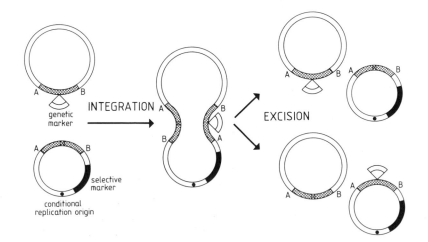

Figure 2. Scheme of the integration-excision step used to exchange allelic sequences between two replicons. One of the two allelic sequences is tagged with a genetic marker whose phenotype can be easily screened for in *E. coli*. Integration is selected at conditions at which one of the two replication origins is not functional. Other details are given in the text.

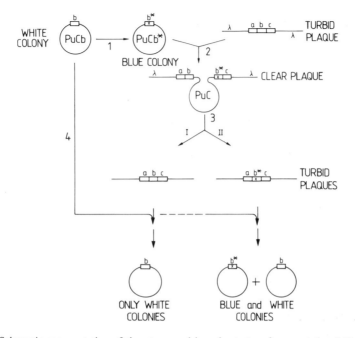

Figure 3. Schematic representation of the steps used in order to transfer a mutation (b*) from a small plasmid to a lambda phage. Circular and linear structures represent plasmids and phages, respectively; a, b and c represent regions of the gene under study. Numbers 1−4 represent steps of the protocol: 1) Selection of a frameshift mutation (in the insert b) which restores the correct reading frame in the gene encoding the α peptide of the enzyme β-galactosidase. Selection is possible because this mutation has the phenotype of blue colonies. 2) Selection of a clear plaque formed by a phage which has the 'mutant' plasmid inserted into its genome. 3) Selection of turbid plaques representing excision of the plasmid. 4) Identification of phages which carry the desired mutation by their ability to donate 'blueness' to the wild-type plasmid.

its replication origin is blocked. Selection for excision can be obtained in different ways depending on the type of replicons and markers used (for a more detailed description of this technology see appendix J in reference 16).

In most cases the allelic sequences to be exchanged between the two replicons do not have a phenotype in *E. coli*. It is therefore impossible to recognise which of the recombinant clones contains the desired allele without lengthy sequence analysis of a large number of clones. Thus, it is convenient to have the DNA region that has to be exchanged labelled with a genetic marker that can be easily recognised in *E. coli*. Any of the markers described above or drug resistance markers can be used for this purpose.

We have recently developed a technique that could be useful when mutations contained in small DNA fragments cloned in plasmids have to be transferred to larger genomic fragments carried in lambda phage vectors. It relies on a method to select double recombination events and on a quick test to identify deletion mutations by exploiting their ability to cause a frameshift alteration in a gene coding for a peptide whose phenotype can be easily tested. As we have already mentioned in Section 1 of this chapter Traboni *et al.* (7) have described the possibility of selecting for frameshift mutants in short sequences inserted into the DNA

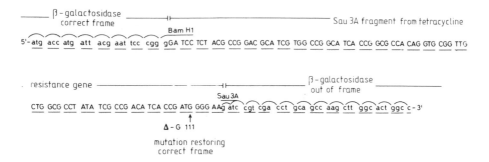

Figure 4. Nucleotide sequence of the 91 base-pair fragment inserted into the polylinker region of pUC8. Lower case letters indicate nucleotides belonging to pUC8, while capital letters indicate those belonging to the 91 base-pairs fragment derived from the tetracycline-resistant (*tet*[R]) gene of pBR322. The alteration of the reading frame after the insertion is shown by the shift from the correct codons, indicated by arcs over the original pUC8 triplets, to the new codons underlined by straight lines. The deletion of one base-pair (ΔGlll), which restores the correct frame is also shown.

Table 5. Integration of Plasmids Carrying Mutations into Phage Carrying the Wild-type Gene.

This is illustrated by the selection of hybrids formed between the phage G6 (λabc in *Figure 3*) and the plasmid pUC9-91/3 f15 (PuCb* in *Figure 3*).

1. Inoculate one colony of *E. coli* strain 71/18 containing the plasmids pUC8-91/3 f15 into 5 ml of L-broth. Grow to saturation by shaking at 37°C.
2. Prepare one L-plate (1% agar in L-broth[a]) containing 0.2 ml of 71/18 (pUC8-91/3 f15) in 4 ml of top agar (0.7% agar in L-broth). Let it set for 10 min at room temperature.
3. Spot $10^5 - 10^7$ plaque forming units of a lysate of phage G6 onto the plate and incubate overnight at 37°C.
4. Pick some top agar from the area of bacterial lysis using a Pasteur pipette and inoculate into 1 ml of phage diluent[b]. Add 1 drop of chloroform.
5. Pipette 100 μl, 30 μl and 10 μl of this phage suspension into three tubes.
6. Add 100 μl of 71/18 in 10 mM $MgSO_4$ to the three tubes and leave for 10 min at room temperature.
7. Add 3 ml of top agar to each tube and plate on L-plates. Incubate at 37°C. Areas of clear lysis should appear on a background of turbid plaques. Clear plaques are derived from the integration of a plasmid into the λ chromosome.
8. Pick a clear plaque using a Pasteur pipette and inoculate it into 1 ml of phage diluent.
9. Dilute this suspension 100-fold with phage diluent and inoculate 100 μl and 20 μl into separate 100 μl aliquots of *E. coli* 71/18 in 10 mM $MgSO_4$.
10. Add 3 ml of top agar to each tube. Incubate overnight at 37°C. This should give a lawn with well isolated clear plaques formed by the hybrid between phage and plasmid.

[a]The composition of L-broth is Bacto-tryptone 10 g, Yeast extract 5 g, NaCl 5 g per liter of solution.
[b]The composition of phage diluant is 10 mM Tris-HCl pH 7.4, 10 mM $MgSO_4.7H_2O$, 0.01% gelatin.

fragment encoding the α-peptide of the enzyme β-galactosidase. We will demonstrate how these mutations can be introduced into their original genetic background by integration and excision of a plasmid carrying the frameshift mutation into a phage in which the whole gene was cloned. However, this method is not confined to frameshift mutations. Once a small deletion which alters the reading frame of the β-galactosidase gene in the tester plasmid has been obtained, any base substitution in the sequence included in the deletion can substitute the deletion itself in the original genetic background. Phages carrying the deletion or the base substitution after the integration excision step, can be recognised by their

Table 6. Selection of Phages that have Excised the Plasmid.

A. *EDTA Selection*

1. Pick ten well isolated, clear plaques formed by the hybrid between phage and plasmid and inoculate into 10 tubes each containing 1 ml of sterile 0.1 M EDTA, pH 8.0. Incubate for 60 min at 50°C.

2. Plate 5 μl and 20 μl of the suspension with 0.2 ml of *E. coli* 71/18 (in 10 mM MgSO₄) and 3 ml of top agar on separate L-plates[a]. Incubate overnight at 37°C. Inspect for turbid plaques[b].

B. *Blue Colony Test for the Mutation*

3. Prepare an L plate containing 200 μl of a saturated culture of 71/18 containing pUC8-91/3 in 4 ml of top agarose (0.7% agarose in L-broth).

4. Using sterile toothpicks pick turbid plaques from the plates after the EDTA selection and streak in an ordered grid onto the agarose plate prepared in step 3. Incubate overnight at 37°C.

5. Replica plate onto an L-plate covered with 4 ml of top agarose containing 200 μl of a saturated culture of 71/18. Cool down the plates by leaving them for about 1 h at 4°C. Then incubate overnight at 37°C.

6. Clear dots should appear inside the turbid replicated streaks. Using a Pasteur pipette pick one clear plaque from each streak and inoculate into 3 ml of L broth containing 500 μg/ml of ampicillin. Grow by shaking overnight at 37°C.

C. *Enrichment for Ampicillin-resistant Colonies Containing Excised Plasmids in a Miniplasmid Preparation*

7. Centrifuge each 3 ml culture from step 6. Suspend each pellet in 1 ml of 10 mM Tris-HCl pH 8.5, 1 mM EDTA and transfer to a microcentrifuge tube. Centrifuge for 2 min in a microcentrifuge.

8. Suspend the pellet in 100 μl of 15% sucrose, 50 mM Tris-HCl pH 8.5, 50 mM EDTA. Add 50 μl of 5 mg/ml lysozyme.

9. Freeze in a bath of dry ice/ethanol and then thaw at room temperature.

10. Leave 15 min in ice. Then add 300 μl of ice-cold water and leave for 5 min on ice.

11. Incubate at 70°C for 15 min, then spin down chromosomal DNA and cell debris for 15 min in a microcentrifuge.

12. Remove the pellet using a hook-ended Pasteur pipette. Add 70 μl of 5 M sodium perchlorate to the supernatant and then 200 μl isopropanol.

13. Centrifuge in a microcentrifuge for 15 min.

14. Redissolve the pellet in 100 μl of 0.3 M sodium acetate. Add 300 μl of ethanol and leave at −70°C for 15 min.

15. Remove the supernatant and wash the pellet with 70% ethanol.

16. Dry the pellet in a vacuum centrifuge. Then redissolve it in 100 μl of water.

17. Add 20 μl of this to 200 μl of 71/18 competent cells prepared as described in the footnote of *Table 4*.

18. Incubate on ice for 30 min and then shift to 42°C for 2 min.

19. Spread the transformed cells onto an L-plate containing 100 μg/ml of ampicillin, 40 μl of X-gal (20 mg/ml in dimethyl formamide) and 20 μl of IPTG (20 mg/ml in water). Incubate overnight at 37°C. The appearance of blue colonies indicates that the original turbid plaques were formed by a phage containing the frameshift mutation.

---

[a]For the composition of L-agar plates see step 2 of *Table 5*.
[b]In this example it is possible to test for resistance to tetracycline. Pick isolated turbid plaques (3 − 5 for each original clear clone) with sterile toothpicks and streak these onto an L-plate and an L-plate containing 5 μg/ml of tetracycline. Clones which grow on L-plates but not on L-plates containing tetracycline are found at a frequency of approximately 5%.

ability or inability to cause a frameshift alteration in the tester plasmid. All the steps in the method (integration, excision, test for the presence of the mutation) are either selective or can be easily screened for.

An outline of the method is shown in *Figure 3*. Although the method was developed to be applicable to DNA fragments which do not have a phenotype in *E. coli*, we will illustrate its use by reference to a gene which confers resistance to the antibiotic tetracycline on the host bacterium. The integration-excision steps occur between a plasmid and a λ phage carrying the substitution i21. The plasmid pUC8-91/3 (PuCb in *Figure 3*) consists of the vector pUC8 with insertion of a fragment of 91 bases obtained by *Sau*3A digestion of the tetracycline resistance gene of pBR322 (17). This insertion of $3N+1$ bases in the gene which encodes the α peptide of β-galactosidase disrupts the translation reading frame and so prevents the formation of an active α-peptide fragment. Plasmid pUC8-91/3 f15 (PvCb* in *Figure 3*) was obtained from pUC8/91/3 by deleting a G residue by partial DNAase treatment (18). This deletion (*Figure 4*, mutation ΔGlll) restores the correct reading frame and as a consequence allows the synthesis of active β-galactosidase and the formation of blue colonies on X-gal indicator plates. The blue colony phenotype is used in the course of the experiment as a marker for the deletion during the various transfers from one background to the other. Phage G6 (λabc in the example of *Figure 3*) is a derivative of NM616 (19) where the *Eco*RI fragment containing the aminoterminal part of the β-galactosidase gene was substituted with the *Pvu*II-*Eco*RI fragment of pBR322 containing the tetracycline resistance gene.

The aim is to exchange allelic sequences by double recombination achieved through an integration excision step. In the example described here, selection for integration takes advantage of the aberrant immunity phenotype which λ i21 acquires when plasmids are integrated into it (20). Because of plasmid replication, the hybrid forms clear instead of turbid plaques and so can be readily identified (*Figure 3*, step 2). A detailed protocol for this step is given in *Table 5*. After integration has been achieved, plaques derived from phages from which the plasmid has been excised (*Figure 3*, step 3) can be enriched by incubation in the presence of chelating agents (21). This procedure is described in *Table 6*. At this stage it is possible to test the phage clones for the successful insertion of the mutation into the complete gene since the experiment makes use of a gene whose phenotype can be readily screened (tetracycline resistance). In a situation where the mutation would have no phenotype in *E. coli*, it is necessary to rely upon a different test. This test is based on the phenotype of the deletion *per se* in *E. coli*, that is, the ability to donate blueness to plasmid pUC8-91/3. In other words phages which contain the deletion can be identified via a second integration excision process with plasmid pUC8-91/3 because they are able to excise plasmids which synthesise an active α-peptide. The protocol for this is contained in *Table 6*.

3. ACKNOWLEDGEMENTS

We thank Wendy Moses for typing this manuscript. The work done in Italy was supported by Progetto Finalizzato Ingegneria Genetica e Basi Molecolari delle Malattie Ereditarie, CNR, Rome, Italy.

4. REFERENCES

1. Fritz,H.-J. (1985) in *DNA Cloning Volume I — A Practical Approach*, Glover,D.M. (ed.), IRL Press Ltd., Oxford, pp. 151-163.

2. Traboni,C., Cortese,R., Ciliberto,G. and Cesareni,G. (1983) *Nucleic Acids Res.*, **11**, 4229.
3. Traboni,C., Ciliberto,G. and Cortese,R. (1984) *Cell*, **36**, 179.
4. Dente,L., Cesareni,G. and Cortese,R. (1983) *Nucleic Acids Res.*, **11**, 1645.
5. Zakour,R.A. and Loeb,L.A. (1982) *Nature*, **295**, 708.
6. Borrias,W.E., Wilshut,J.T.C., Vereijken,J.M., Weisbeck,P.J. and Van Arkel,G.A. (1976) *Virology*, **70**, 195.
7. Traboni,C., Ciliberto,G. and Cortese,R. (1982) *EMBO J.*, **1**, 415.
8. Ryan,H.J., Belagaye,R., Brown,E.L., Fritz,H.J. and Khorana,H.G. (1979) *J. Biol. Chem.*, **254**, 10803.
9. Shortle,D. and Nathans,D. (1979) *J. Mol. Biol.*, **131**, 801.
10. Everett,R.D. and Chambon,P. (1982) *EMBO J.*, **1**, 433.
11. Folk,W.R. and Hoffstatter,H. (1983) *Cell*, **33**, 585.
12. Ciampi,M.S., Melton,D.A. and Cortese,R. (1982) *Proc. Natl. Acad. Sci. USA*, **79**, 1388.
13. Wieriga,B., Meyer,F., Reiser,J. and Weissmann,C. (1983) *Nature*, **301**, 38.
14. Shortle,D., Grisafi,P., Benkovic,S.J. and Botstein,D. (1982) *Proc. Natl. Acad. Sci. USA*, **79**, 1588.
15. Kramer,W., Schughart,K. and Fritz,H.J. (1982) *Nucleic Acids Res.*, **10**, 6475.
16. Silhavy,T.J., Berman,M.L. and Enquist,L.W. (1984) *Experiments with Gene Fusions*, Cold Spring Harbor Laboratory Press, Cold Spring Harbor, NY.
17. Vieira,J. and Messing,J. (1982) *Gene*, **19**, 259-268.
18. Lorenzetti,R., Cesareni,G. and Cortese,R. (1983) *Mol. Gen. Genet.*, **192**, 515.
19. Mileham,A.J., Revel,H.R. and Murray,N.E. (1980) *Mol. Gen. Genet.*, **197**, 227.
20. Windass,J.D. and Brammar,W.J. (1979) *Mol. Gen. Genet.*, **172**, 329.
21. Parkinson,J.S. and Davis,R.W. (1971), *J. Mol. Biol.*, **56**, 425.
22. Messing,J., Groneborn,B., Muller-Hill,B. and Hofschneider,P.H. (1977) *Proc. Natl. Acad. Sci. USA*, **74**, 3642.

CHAPTER 8

The Oligonucleotide-directed Construction of Mutations in Recombinant Filamentous Phage

HANS-JOACHIM FRITZ

1. INTRODUCTION AND THEORETICAL BACKGROUND

1.1 Directed Mutagenesis and the Construction of Precise Mutations

Directed mutagenesis and related techniques have recently attracted considerable attention. The reason for this is that these methods can generate mutations with frequencies in the percent range. This makes it possible to perform genetic analyses in the absence of a recognisable phenotype. The mutations can, if necessary, be detected and characterised by biochemical methods alone and the conventional path of a genetic analysis − from the phenotype 'inward' to the gene − can be reversed. This is important since mutants without an acccessible phenotype are in many cases equally or even more interesting than ones accessible as the result of spontaneous events or by conventional mutagenesis.

The methods can be divided into two broad categories:

(i)   Methods to saturate a more or less precisely defined target region with (single) point mutations.
(ii)   Methods to construct precise, predetermined, and possibly complex, structural changes within the target region.

The two types of methods, rather than being alternatives, each have their own field of application. The choice of one or the other depends on the nature of the question to be asked in a particular experiment.

Among the methods in the second category, the most frequently used is that of marker salvage (1) from synthetic oligonucleotides (2). The DNA fragment to be mutated is cloned in a suitable vector − most commonly a derivative of filamentous phage (like members of the M13mp series). Single stranded virion DNA of the recombinant phage is isolated as a target for mutagenesis. A short segment (usually about 15 nucleotides) of the complementary strand is synthesised containing the desired structural change. The synthetic oligonucleotide is then specifically annealed to the target DNA site and used as a primer in the enzymatic *in vitro* synthesis of fully double-stranded DNA. The resulting DNA molecule is complementary throughout, with the exception of the site to be mutated. This DNA heteroduplex is then used for transfection of *E. coli* cells and the phage progeny obtained is screened for descendents of the (−) strand of the transfecting

DNA heteroduplex, since this strand contained the synthetic oligonucleotide.

This type of mutagenesis is extremely flexible with respect to the nature of the structural change that can be induced. Point mutations, deletions and insertions have all been constructed. If a pure, specific primer is used, then experiments of the second category result in a single well-defined mutation. This is in contrast to experiments that produce mutant collections, which fall into category (i).

The major advantages of the method are:

(i)   Working hypotheses concerning structure/function relationships in biological macromolecules can be tested stringently by very straightforward experiments.

(ii)   Once the understanding of this structure-function relationship for a given biological macromolecule has evolved to a sufficiently high degree, the same method can be employed to alter the properties of this molecule in a pre-determined fashion. The first examples of this approach have already been described (3).

1.2 The Essential Biology of Phage M13

The oligonucleotide-directed construction of mutations requires single stranded DNA target regions for hybridisation of the mutagenic primer. It is, therefore, most convenient to clone the DNA region to be modified into a derivative of phage M13. This provides a system for 'biological strand separation'. The most salient features of the M13 life cycle are summarised in a chapter of the textbook *'DNA replication'* (18). Phage M13 was converted to a cloning vehicle by introducing a short fragment of *E. coli lac* DNA into its genome. Subsequently a number of restriction sites were introduced into this *lac* DNA fragment (4). Phage M13 infects *E. coli via* attachment to sex pili. Thus, only F^+ and Hfr strains are suitable for work with M13 cloning vehicles.

1.3 Possible Mechanisms for Mutation Fixation

After entering the host cell, the transfecting heteroduplex DNA can have two kinds of fate other than the possible loss of one or the other strand:

(i)   The two markers can segregate passively by semiconservative replication. This will lead to a mixed burst and marker yields up to 50% can be expected for θ-type replication. For phage M13, a bias in favour of the synthetic marker is to be expected due to the asymmetric replication of its genome.

(ii)   The heteroduplex DNA is subject to mismatch repair prior to the onset of replication. In this case, a pure burst will result and the marker yield depends on the bias with which the DNA mismatch repair machinery chooses the template strand. It has been demonstrated that this bias has indeed a great impact on the construction of mutations (5) (also see below).

1.4 DNA Mismatch Repair in Escherichia coli and the Importance of Adenine Methylation in the Transfecting Heteroduplex DNA

A DNA mismatch repair system apparently operates in *E. coli* contributing to the overall fidelity of DNA replication by the post-replicative correction of nucleotide misincorporations that have escaped proofreading (6). The strand choice in this repair process is most likely directed by transient undermethylation of GATC sites in the newly synthesised strand $(7-9)$. In other words, the post-replicative DNA mismatch repair system recognises the parental information to be conserved in the methylated DNA strand and repairs the deviant structure in the unmethylated strand.

The sequence GATC is the target site of adenine-N^6 methylation by the *E. coli dam* methylase (10). These sites are under-represented in the genome of filamentous phage: three sites exist in the genome of M13 (11) and four in the genomes of both phages f1 and fd (12). Nevertheless, it was shown that at least certain types of single nucleotide mismatches in artificially constructed heteroduplex DNA molecules of these and related phage genomes are subject to methyl-directed mismatch repair both *in vivo* (5,13) and *in vitro* (14). This has a number of consequences for oligonucleotide-directed mutation. The most frequently used protocol to date (2) employs the following sequence of steps:

(i)   Isolation of (methylated) M13 virion DNA from a wild type (i.e., *dam*$^+$) host.

(ii)   Annealing of the synthetic oligonucleotide to the recombinant phage genome.

(iii)   *In vitro* fill-in reaction with DNA polymerase I, accompanied by sealing the resulting double stranded DNA with DNA ligase.

(iv)   Biochemical enrichment for covalently closed circular DNA.

(v)   Transfection.

Obviously, the transfecting heteroduplex DNA is now hemimethylated with the methylated adenine residues being part of the (+) strand. Thus, if the unpaired DNA site is susceptible to DNA mismatch repair, the repair activity will be directed *against* the desired synthetic marker. It is therefore not surprising that the conventional protocol typically gives rise to marker yields of only about 5%. Here we describe a method that circumvents this problem and, on the contrary, exploits the methyl-instructed DNA mismatch repair system of the recipient cell by using hemimethylated heteroduplex DNA with the methyl groups in the (−) strand, for transfection.

1.5 The Gapped Duplex DNA Approach Employing Heteroduplex DNA with Properly Orientated Hemimethylated GATC Sites

Hemimethylated heteroduplex DNA with the methylated adenine residues in the (−) strand is constructed as follows:

(i)   The DNA fragment to be mutated is cloned into an M13 vector and unmethylated virion DNA is isolated after propagation of the recombinant phage in a *dam*$^-$ host strain.

(ii)   Methylated RF-DNA of the M13 *vector* alone is isolated after propagation in a *dam*$^+$ host.

(iii)   The RF-DNA is cleaved at the restricton site(s) that was(were) used for cloning the insert DNA.

(iv)   The linearised RF-DNA is heat-denatured and renatured in the presence of an excess of recombinant virion DNA. The result is a hemimethylated partial DNA duplex (gapped duplex DNA, gdDNA) with the methyl groups in the incomplete ($-$) strand. The exogenous DNA in the recombinant phage genome is exposed in single stranded form.

(v) The mutagenic primer is annealed to the single-stranded gap, which is then filled-in using DNA polymerase and DNA ligase as mentioned above.

(vi) The molecule is introduced into *E. coli* by transfection and the mutant and wild-type strands segregate during replication.

(vii) The resulting phage progeny are used for re-infection at low multiplicity.

The advantages of the method are three-fold:

(i) Annealing of the mutagenic primer to inappropriate sites is suppressed since only a relatively small window of single stranded DNA is available for hybridisation.

(ii) The gap-filling reaction is facilitated since in the gdDNA, the entire vector part of the recombinant phage genome is already covered by the complementary ($-$) strand.

(iii) Mismatch repair acting on the unpaired DNA site is biased in favour of the synthetic marker.

Marker yields around 45% have been obtained reproducibly with this method *without* enriching for covalently closed circular DNA for transfection. These yields are high enough to make screening of large numbers of clones (e.g., by selective 'dot-blot' hybridisation) unnecessary. Direct DNA sequence analysis performed on a small number of candidate clones is now the most economical way of screening in the absence of an easily discernible phenotype.

1.6 Limitations

The oligonucleotide-directed construction of mutations as described here has a number of general and specific limitations. The following are of special practical importance:

(i) Redundant DNA sequences present in the target region for mutation can cause problems with primer annealing. Direct repeats will result in equivocal choice of the primer hybridisation site. Inverted repeats can form snap-back structures and prevent primer annealing altogether.

(ii) Not all unpaired DNA sites are equally subject to methyl-directed DNA mismatch repair (13,15). Thus, while the method described here will always prevent the occurrence of unfavourably oriented DNA mismatch repair, one cannot expect the added advantage (over passive segregation) of favourably oriented repair to be realised in every individual case.

2. EXPERIMENTAL APPROACHES

2.1 The Synthetic Primer

It is mandatory that the synthetic oligonucleotide used for mutation construction be of the highest purity and be quantitatively phosphorylated at the 5'-terminus. Such oligonucleotides may be synthesised by a number of chemical methods and are also available from several commercial suppliers. A volume of the *Practical*

Table 1. Preparation of Unmethylated Recombinant Phage Virion DNA (ssDNA).

1.	Prepare a 2–3 ml overnight culture of *E. coli* strain GM99-F2su$_{III}$[a] at 32°C in 'Antibiotic Medium 3' (Difco).
2.	Inoculate 200 ml 'Antiboitc Medium 3' in a 1 l Erlenmeyer flask with F2su$_{III}$ and 0.1–0.2 ml phage suspension[b] with a titre of about 10^{12} p.f.u./ml[c]. Shake overnight at 32°C.
3.	Withdraw a 1.5 ml aliquot into a microcentrifuge tube. Use for verification of the Dam$^-$ phenotype (*Table 2*) and determination of phage titre (*Table 3*).
4.	Transfer the rest of the culture into a 250 ml plastic screw cap centrifuge bottle and centrifuge for 30 min at 10 000 r.p.m. in a Sorvall GSA-rotor or its equivalent between 4°C and 15°C.
5.	Pour the supernatant into another plastic screw cap centrifuge bottle and add 8 g PEG 6000 (poly-ethylenegylcol) and 6 g NaCl (final concentration ~0.5 M). Dissolve the PEG and NaCl and keep the mixture at 4°C for at least 4 h to precipitate the phage[d].
6.	Pellet the precipitated phages by centrifugation at 10 000 r.p.m. for 20–30 min at 4°C in a Sorvall GSA-rotor or its equivalent. Pour off the supernatant carefully and allow any residual liquid to drain from the bottle onto paper towels for about 10 min.
7.	Resuspend the phage pellet in 20 ml of 10 mM Tris-HCl pH 8.0, 50 mM NaCl, 1 mM EDTA using a 10 ml pipette. Transfer the suspension to a 40 ml plastic screw cap centrifuge tube. Wash the bottle with another 10 ml of the same buffer and add the wash to the other suspension.
8.	Add 0.6 g PEG 6000 and 0.9 g NaCl to the 30 ml of suspension. Dissolve the PEG and NaCl and keep the mixture at 4°C for at least 1 h.
9.	Pellet the precipitated phages by centrifugation at 15 000 r.p.m. for 10 min in a Sorvall SS34-rotor or its equivalent at 4°C. Pour off the supernatant carefully and allow the tubes to drain on paper towels for about 10 min.
10.	Resuspend the phage pellet with a 1 ml pipette in 2 ml of 10 mM Tris-HCl pH 8.0, 50 mM NaCl, 1 mM EDTA and transfer the suspension to a 14 ml capped polypropylene tube. Wash the centrifuge tube with 1 ml of the same buffer and transfer the wash to the same capped tube.
11.	Add 3 ml of phenol, saturated with 10 mM Tris-HCl pH 8.0, 0.2 mM EDTA, vortex extensively and centrifuge for about 10 min at 6000 r.p.m. in a bench-top centrifuge at room temperature. Transfer the aqueous (upper) phase to a fresh tube, taking care not to transfer any of the interphase.
12.	Repeat the extraction of the aqueous phase with an equal volume of a 1:1 mixture (v:v) of phenol and chloroform.
13.	Extract the aqueous phase three times with diethylether. Remove the final ether phase and allow the remaining ether to evaporate at 65°C for 15–30 min[e].
14.	Determine the volume of the aqueous phase and add 1/10 volume 3 M sodium acetate and three volumes ethanol (−20°C). Store at −20°C for at least 1 h.
15.	Pellet the DNA by centrifugation at 15 000 r.p.m. for 20 min in a Sorvall SS34-rotor or its equivalent at 0°C. Carefully pour off the supernatant and allow the tube to drain for about 10 min on a paper towel. Dry the pellet in a vacuum desiccator.
16.	Dissolve the pellet in 2 ml of 10 mM Tris-HCl pH 8.0, 0.25 mM EDTA and determine the yield[f]. Store the DNA at +4°C or −20°C.

[a]The genotype of *E. coli* GM99-F2su$_{III}$ is *dam-4, mal354, tsx, su$_{III}$(ϕ80); F'gal/T::Tn5 (kan^r)*.
[b]Phage stocks are grown in *E. coli* BMH 71-18 ($\triangle$(*lac-pro*), *thi, supE; F'lac, Z$\triangle$M15, proA$^+$B$^+$)*).
[c]See *Table 3* for instructions for the titration of phage stocks.
[d]This is a convenient point at which to verify the Dam$^-$ phenotype (*Table 2*) and to titre the phage (*Table 3*).
[e]Do not allow the ether near naked flames or electrical switches.
[f]A typial yield is about 300–1000 μg/200 ml culture. The yield can be lower for phages with amber mutations such as M13mp8 and M13mp9.

Approach series of books that is entirely devoted to the preparative and analytical aspects of chemical oligonucleotide synthesis is available (19).

On an empirical basis, the following structural requirements for the mutagenic

Table 2. Verification of Dam⁻ Phenotype.

1.	Centrifuge the 1.5 ml suspension of cells and phages from Step 3 of *Table 1* for 1 min in a microcentrifuge. Remove the supernatant carefully with a Pasteur pipette and save it for phage titration (*Table 3*).
2.	Resuspend the cell pellet by flicking or vortexing, with 100 µl of an ice-cold solution of 50 mM glucose, 10 mM EDTA, 25 mM Tris-HCl pH 8.0 to which 4 mg/ml lysozyme has been added just before use. Incubate for 5 min at room temperature[a].
3.	Add 200 µl of freshly prepared 0.2 N NaOH, 1% SDS and mix by inverting the tube 2−3 times.
4.	Keep the tube in ice for 5 min, and then add 150 µl ice-cold solution of 3 M potassium acetate pH 4.8 (adjusted with 98−100% acetic acid). Gently vortex the tube in an inverted position and keep on ice for a further 5 min.
5.	Centrifuge the tube for 5 min at 4°C in a microcentrifuge.
6.	Transfer the supernatant to a fresh microcentrifuge tube and add to it an equal volume of a 1:1 mixture (v:v) of phenol and chloroform. Vortex the mixture and then centrifuge the tube for 2 min in a microcentrifuge.
7.	Transfer the supernatant to a fresh tube and add two volumes of ethanol and 0.1 volumes of 3 M sodium acetate. Mix, and leave at room temperature for 2 min.
8.	Centrifuge the tube for 5 min in a microcentrifuge. Remove the supernatant carefully using a drawn out Pasteur pipette. Add 1 ml 70% ethanol to the pellet, vortex and centrifuge again for 5 min in a microcentrifuge.
9.	Remove the supernatant as before and dry the pellet in a Speed-Vac-Concentrator or a vacuum desiccator.
10.	Dissolve the pellet in 20 µl of 10 mM Tris-HCl pH 8.0, 0.25 mM EDTA containing 20 µg/ml of DNase-free pancreatic RNase. Add 3 µl of 500 mM NaCl, 100 mM Tris-HCl pH 7.5, 100 mM MgCl₂, 10 mM DTT and 6 µl water. Add 1 µl *Mbo*I (10 u/µl) and incubate for 1 h at 37°C.
11.	Analyse the DNA for cleavage by *Mbo*I by gel eletrophoresis on 1−1.5% agarose. Lack of cleavage can be due to any of the three following reasons: (i) reversal of the host strain to a Dam⁺ phenotype, (ii) an inactive restriction endonuclease, (iii) a poor DNA preparation.

[a]This and subsequent steps to prepare DNA are a variation of the procedure developed by Birnboim and Doly described in pp 368−369 of reference 20.

Table 3. Determination of Phage Titre.

1.	Dilute the supernatant from Step 1 in *Table 2*, to 10^{-3}, 10^{-6} and 10^{-9} with 20 mM sodium phosphate pH 7.2, 0.002% gelatine.
2.	Place four sterile 12 ml glass tubes containing 2.5 ml of thoroughly melted top agar in a 45°C water bath. When the agar is cooled to 45°C, add two drops of a fresh overnight culture of *E. coli* BMH 71-18[b] from a sterile 1 ml pipette to each tube.
3.	Add 100 µl of the various phage dilutions to three of the tubes. Mix by vortexing and pour the top agar onto EHA-plates[c]. The fourth tube without phages is also plated as a control to determine whether the strain is contaminated.
4.	Allow about 20 min at room temperature for the top agar to solidify, and then incubate the plates upside down at 37°C for at least 6 h.
5.	Count the plaques and calculate the phage titre. It is only worth continuing a preparation of virion DNA (*Table 1*) if the titre is at least 5 x 10^{11} p.f.u./ml. If phages contain amber codons in vital genes, such as M13mp8 and M13mp9, then a somewhat lower titre is acceptable.

[a]Top agar contains 10 g of Bacto-tryptone (Difco), 5 g NaCl, 6.5 g Bacto-agar (Difco) and 1 ml of 1 M MgSO₄ per litre of water.
[b]The genotype of *E. coli* BMH 71-18 is given in footnote [b] of *Table 1*.
[c]EHA-agar is made from 13 g Baco-tryptone (Difco), 8 g NaCl, 2 g sodium citrate-dihydrate, and 10 g Bacto-agar per litre of water. The mixture is autoclaved and 7 ml of sterile 20% glucose is then added. The medium should be poured onto 9 cm Petri dishes several days before use.

Table 4. Preparation of Methylated Vector RF-DNA.

1.	Prepare a 75 ml overnight culture of *E. coli* BMH 71-18[a] in 'Antibiotic Medium 3' (Difco) at 37°C. Dispense 3 l of 'Antibiotic Medium 3' into three sterile 2 l Erlenmeyer flasks. Inoculate each flask with 25 ml of the overnight culture of BMH 71-18 and shake at 37°C until the culture has reached an O.D.$_{546}$ of 0.5−0.6.
2.	Infect each culture with 25 ml of a phage suspension with a titre of about 10^{12} p.f.u./ml and continue shaking at 37°C for another 3−4 h.
3.	Pellet the cells by centrifugation. Pour off the supernatant and resuspend the cells at 0°C in 22.5 ml 25% sucrose/50 mM Tris-HCl pH 8.0 using a 10 ml pipette. If several bottles were used for centrifugation, add the 22.5 ml to the first bottle, resuspend the cells, transfer the suspension to the next bottle and so on.
4.	Add 6 ml lysozyme solution (5 mg/ml in 50 mM Tris-HCl pH 8.0) and shake gently for 10 min at 0°C.
5.	Add 7.5 ml of 250 mM EDTA pH 8.0 and continue gentle agitation at 0°C for 10 min.
6.	Add 27.5 ml 50 mM Tris-HCl pH 8.0, 1% Brij 58[c], 0.4% sodium deoxycholate, 62.5 mM EDTA and again shake gently at 0°C for 10 min.
7.	Carefully transfer the mixture, which should now be viscous and cloudy, to screw cap tubes for the Beckman type-30 rotor or its equivalent and centrifuge at 30 000 r.p.m. for 60 min at 4°C.
8.	Pool the supernatants in a 100 ml measuring cylinder, being careful not to transfer parts of the viscous pellets. Add 0.97 g CsCl and 20 μl of a 10 mg/ml solution of ethidium bromide per ml of supernatant.
9.	Distribute the solution equally into two 35 ml polyallomer 'quick-seal' centrifuge tubes. Fill up the tube with taring-solution[d]. Seal the tubes following the manufacturer's instructions and centrifuge for at least 14 h at 45 000 r.p.m., 15°C in a Beckman VTi50 rotor or its equivalent.
10.	After centrifugation two (sometimes three) bands should be visible, corresponding to linear and nicked-circular DNA and to supercoiled DNA (the potential third band is single-stranded DNA). Stick a syringe needle into the top of the tube in order to allow air to enter the tube when the bands are collected. Pierce the wall of the tube with a needle tipped syringe just below the upper band(s) and draw off these bands. Without removing the syringe, insert another needle tipped syringe into the tube just below the lowest band containing the supercoiled DNA and draw off this band in a volume of 5 ml per tube.
11.	Transfer these 5 ml samples to two 5 ml 'quick-seal' tubes and re-centrifuge for at least 8 h at 45 000 r.p.m. in a Beckman VTi65 rotor or its equivalent at 15°C.
12.	Collect the bands of supercoiled DNA as described in step 10 each in a volume of 1−2 ml and dialyse against 1000 volumes of 10 mM Tris-HCl pH 8.0, 50 mM NaCl, 1 mM EDTA.
13.	Transfer the DNA solution into a 14 ml capped polypropylene tube, and add an equal volume of phenol saturated with 10 mM Tris-HCl pH 8.0, 0.25 mM EDTA. Mix by shaking and separate the phases by centrifugation. Transfer the aqueous (upper) phase to a new tube and repeat the extraction another two times.
14.	Extract the aqueous phase three times with two volumes of diethylether. Remove the final ether phase and warm the aqueous phase to 65°C for 15−30 min[e]. Dialyse 2−3 times against 1000 volumes of 10 mM Tris-HCl pH 8.0, 50 mM NaCl, 1 mM EDTA.
15.	Determine the DNA yield[f] photometrically using the last dialysis buffer as reference. Check that the A_{260}/A_{280} absorbance ratio is close to 2.0. Store the DNA at −20°C.

[a]*E. coli* DMH 71-18 is described in footnote[b] of *Table 1*.
[b]The phage suspension should be prepared following steps 1−4 of *Table 1*.
[c]Brij 58 is polyethylenegycolmonostearylether.
[d]Taring solution is 50 mM Tris-HCl pH 8.0 to each 1 ml of which has been added 0.97 g CsCl and 20 μl of 10 mg/ml ethidium bromide solution.
[e]See footnote[e] to *Table 1*.
[f]The yield should be between 100 and 400 μg of RF DNA per litre of culture.

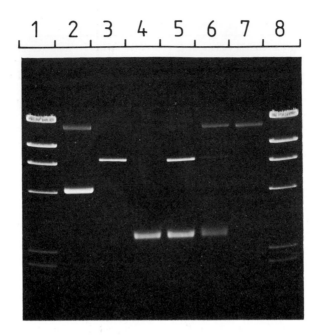

Figure 1. Gel Electrophoretic Analysis of Different Intermediates of the Construction of gdDNA. **Lanes 1** and **8**, λDNA cut with *Hind*III; **lane 2**, RF-DNA of M13mp2; **lane 3**, shortened linearised RF-DNA of M13mp2 (cut with *Eco*RI and *Pvu*I); **lane 4**, single-stranded DNA of M13mp2; **lane 5**, mixture of DNAs shown in **lanes 3** and **4**; **lane 6**, the same DNA mixture as in **lane 5** but after denaturation and renaturation; **lane 7**, purified gdDNA. The DNA preparations and electrophoresis were carried out as described in *Table 5*.

primer can be defined:

(i)   Length of primer: the mutagenic primer consists of a core sequence to be left unpaired after annealing to the M13 genome and two flanking sequences providing the ability of stable and site-specific hybridisation. The flanking sequences should each be about 6 − 7 nucleotides long for a sequence of normal G+C content and an unpaired core of just one nucleotide. If the flanking sequences have a high A+T content and/or if extended unpaired regions must be accommodated in the center, the flanking sequences should be longer.

(ii)   Sequences near the termini of the primer: fraying ends of the primer annealed to the template seem to interfere with the DNA polymerase/DNA ligase reaction. In particular, circumstantial evidence obtained in the authors's laboratory suggests that strand displacement at the 5'-terminus of the primer occurs easily and is a detrimental event (B.Kramer, W.Kramer and H.-J.Fritz, unpublished). Whenever possible, the sequence at the two primer termini should therefore consist of one or several G or C residues.

2.2 Preparation of Gapped Duplex DNA

Starting materials for the preparation of hemimethylated gdDNA are unmethylated virion DNA and methylated RF-DNA of phage M13. Their preparation has been

Table 5. Construction of Gapped Duplex DNA (gdDNA).

1.	Cleave 5 μg of the methylated RF-DNA of the M13 vector with the restriction enzyme(s), used for the cloning of the DNA fragment that is to be mutated. Check that the conversion of RF-DNA to the linear form is compared by electrophoresing a small aliquot on a 0.8% agarose gel[a].
2.	Extract the digestion mixture two times with phenol, saturated with TE buffer[b] four times with diethylether, precipitate the DNA with ethanol and pellet the DNA by centrifugation[c].
3.	Redissolve the precipitate of linearised DNA in TE-buffer to a final concentration of 0.5 μg DNA/ml.
4.	Mix 5 μg (1 pmol) of the linearised RF-DNA with 12.5 μg (5 pmol) of the unmethylated ssDNA of the recombinant phage in a 1.5 ml microcentrifuge tube. Add 7 μl of 10 x SSC[d] and adjust the volume to 70 μl with water.
5.	Transfer the tube to a boiling water bath for 7 min. A prewarmed metal block can be placed over the tube to secure the cap. Transfer the tube into a 65°C bath for 10 min.
6.	Transfer the tube to 0°C, add 30 μl of dye-mix[e] and vortex the mixture for 10 sec. Apply the mixture to a 'long slot'[f] in a 1% agarose gel. Apply 250 ng of RF-DNA as a marker in a small separate slot and electrophorese for 16 h at 30 V[g]. Stain the gel in an aqueous solution of 1 μg/ml ethidium bromide. View under u.v. illumination. Three bands should resolve from the hybridisation mixture corresponding to single-stranded DNA, linearised RF-DNA and, the slowest, to gdDNA.
7.	Cut out the gel piece which contains the band corresponding to the gdDNA and which comigrates with the nicked-circular DNA of the RF-DNA marker. Cut a well into the gel which is somewhat larger than the gel piece and line it with dialysis membrane. Put the gel piece into this well and add electrophoresis buffer until the gel piece is just submerged. Electrophorese at 150 − 200 V until no stained material is left in the gel piece. This normally takes about 30 min. Reverse the orientation of the field for 30 sec.
8.	Remove the buffer containing the gdDNA from the well and, if necessary, reduce the volume to 400 − 500 μl by shaking with n-butanol or by lyophilisation.
9.	Extract the DNA-solution two times with TE-saturated phenol, once with phenol/chloroform (1:1/v:v) and four times with diethylether. Evaporate the remaining ether at 65°C[h].
10.	Precipitate the DNA two times with ethanol[i]. Redissolve the pellet of precipitated DNA in 20 μl TE-buffer.
11.	Take a 1 μl aliquot and apply it to a 1% agarose gel. Apply 1 μg, 500 ng, 250 ng, 125 ng and 62.5 ng of *Hind*III digest of λDNA and 250 ng of RF DNA to the same gel. Electrophorese for 2 − 4 h at 60 V. Estimate the yield by comparison of the intensity of the gdDNA band with that of the 9.3 kb band of the λDNA/*Hind*III digest. The 9.3 kb band contains 1/5 of the total λDNA. Store the gdDNA solution at −20°C.

[a]An example of such a gel is shown in *Figure 1*.
[b]TE is 10 mM Tris-HCl pH 8.0, 0.25 mM EDTA.
[c]See steps 13 − 14 of *Table 1*.
[d]10 x SSC is 1.5 M NaCl, 0.15 M sodium citrate.
[e]Dye-mix is 60% sucrose, 0.05% bromophenol blue, 0.05% xylene cyanol FF, 90 mM Tris-borate pH 8.3, 2.5 mM EDTA.
[f]The gel dimensions are 123 mm x 82 mm x 5 mm and the 'long slot' is 1.4 mm x 31 mm x 3.5 mm.
[g]Electorphoresis buffer is 40 mM Tris-acetate pH 7.4, 5 mM sodium acetate, 1 mM EDTA.
[h]See footnote[e] of *Table 1*.
[i]See steps 7 − 9 of *Table 2*.

described (4). *Tables 1* to *4* give slightly modified versions of these procedures.

With these starting materials at hand, gdDNA can be prepared as described in *Table 5*. *Figure 1* shows an electrophoretic analysis of the different intermediates in the construction of gdDNA. Note that the gdDNA migrates slower in the gel than either of the two starting DNA components.

Table 6. Hybridisation of the Mutagenic Primer to the gdDNA and the Polymerase/Ligase Reaction.

1.	Mix 20 fmol (100 ng) gdDNA and 400 fmol of the mutagenic primer in a 1.5 ml microcentrifuge tube. Add 2 μl of 5x hybridisation buffer[a] and adjust the volume to 10 μl with water.
2.	Heat the mixture to 65°C for 7 min and keep the cap of the tube closed with a prewarmed metal block placed over the tube.
3.	Incubate for 5−15 min at room temperature, then keep at 0°C.
4.	Add 23 μl of water and 4 μl 10x 'fill-in' buffer[b]. Mix and add 2 μl T4-DNA ligase (2 u/μl)[c] and 1 μl DNA polymerase, large fragment (0.2 u/μl)[d].
5.	Incubate the mixture for 45 min at room temperature and then stop the reaction by adding 1 μl 0.5 M EDTA and heating at 65°C for 10 min.
6.	Extract the mixture once with 40 μl phenol, saturated with TE[e] and three times with diethylether; evaporate the residual ether at 65°C[f].
7.	Store the mixture on ice until it is used for transfection (see *Table 7*, steps 3−7). Avoid prolonged storage.

[a]Hybridisation buffer is 750 mM KCl, 50 mM Tris-HCl pH 7.5.
[b]10x 'fill-in' buffer is 625 mM KCl, 275 mM Tris-HCl pH 7.5, 150 mM MgCl$_2$, 20 mM DTT, 0.5 mM ATP and 0.25 mM of each of the four deoxyribonucleoside triphosphates.
[c]1 unit of T4-DNA ligase is the amount of enzyme required to give 50% ligation of *Hind*III digested lambda DNA in 30 min at 16°C at a 5′ termini concentration of 0.12 mM.
[d]1 unit of DNA polymerase catalyses the incorporation of 10 nmol of nucleotide in 30 min using poly dAT as a primer.
The enzyme can be diluted in 50 mM potassium phosphate buffer pH 7.2, 50% glycerol.
[e]See footnote [b] to *Table 5*.
[f]See footnote [e] to *Table 1*.

Table 7. Transfection and Segregation.

1.	Inoculate 50 ml of 'Antibiotic medium 3' (Difco) with a 1 ml overnight culture of *E. coli* BMH 71-18[a]. Grow the culture with shaking at 37°C to an O.D.$_{546}$ of 0.6.
2.	Pellet the cells by centrifugation at 6000 r.p.m. in the Sorvall SS34 rotor or its equivalent at 4°C. Resuspend the cells in 20 ml of ice-cold 100 mM CaCl$_2$ and re-centrifuge. Resuspend the cell pellet in 10 ml of 100 mM CaCl$_2$. Pellet the cells once more by centrifugation and resuspend them in 2 ml of ice-cold 100 mM CaCl$_2$. Keep the cells on ice for at least 30 min.
3.	Add 60 μl of ice-cold 100 mM CaCl$_2$ to the polymerase/ligase reaction mixture (*Table 6*, step 7) and then add 200 μl of the cell suspension from the previous step. Keep the mixture on ice for 90 min.
4.	Vortex briefly and incubate the suspension at 45°C for 3 min.
5.	Take 20 μl of this suspension and prepare 10^{-1} and 10^{-2} dilutions in 100 mM CaCl$_2$. Use 100 μl of each of these dilutions to determine the transfection efficiency as described in *Table 3* but incubate the plates overnight.
6.	Inoculate 25 ml of 'Antibiotic medium 3' with the rest of the transfection mixture (280 μl) and shake overnight at 37°C.
7.	Transfer a 1.5 ml aliquot of the culture to a microcentrifuge tube and centrifuge for 5 min. Save the supernatant and determine the phage titre (see *Table 3*). Store the supernatant at 4°C.

[a]*E. coli* BMH 71-18 is described in footnote [b] of *Table 1*.

2.3 The Construction of Mutations

Starting from hemimethylated gdDNA, the mutation construction protocol consists of four steps:
(i)   Annealing the mutagenic primer to the gdDNA.
(ii)   Gap-filling and sealing with DNA polymerase and DNA ligase.
(iii)   Transfection of an *E. coli* host and preparation of a mixed phage progeny.

Table 8. Preparation of Template for DNA Sequence Analysis.

1.	Inoculate 100 ml of 'Antibiotic medium 3' with a 1 ml overnight culture of *E. coli* BMH 71-18[a]. Dispense 2.5 ml aliquots into sterile 15 ml culture tubes.
2.	Pick at random a number of plaques resulting from Step 7 of *Table 7*. Use small sterile glass tubes such as 200 μl capillary pipettes to pick the plaques. Blow the plaques into the culture tubes.
3.	Incubate with shaking for 5 h at 37°C. Transfer 1.5 ml of each culture to a microcentrifuge tube and centrifuge for 5 min in a microcentrifuge. Save the rest of the culture (see Step 13).
4.	Take out 1 ml of the supernatant with a Gilson or Eppendorf pipette or the equivalent. To avoid transferring cells, slide the pipette tip down along the wall of the tube opposite to the cell pellet.
5.	Centrifuge the supernatant again for 5 min. Transfer 800 μl of the supernatant to another tube following the precautions described in step 4.
6.	Add 200 μl of 2.5 M NaCl, 20% PEG 6000 to the 800 μl of supernatant. Mix and leave for 15 min at room temperature.
7.	Centrifuge for 5 min and remove the supernatant carefully with a Pasteur pipette. Wipe the mouth of the tube with a paper towel. Centrifuge for 2 min and remove all remaining traces of the supernatant with a drawn-out capillary pipette.
8.	Add 110 μl TE[b] to the viral pellet which should be visible. Add 50 μl of TE-saturated phenol and vortex for 15−20 sec. Shake the tubes for 10 min at room temperature on an Eppendorf mixer, and then vortex for 15 sec. Centrifuge for 3 min, and transfer the aqueous (upper) phase to a fresh tube.
9.	Add 100 μl of chloroform, vortex and centrifuge for 2 min. Transfer 90 μl of the aqueous (upper) phase to a fresh tube.
10.	Add 10 μl of 3M sodium acetate and 300 μl of ethanol from a stock kept at −20°C and keep the mixture at −20°C for at least 1 h.
11.	Centrifuge the tube for 15 min at 15 000 r.p.m. in the Sorvall SS34 rotor or its equivalent at 4°C. Take off the supernatant with a drawn out Pasteur pipette. Add 1 ml of cold ethanol to the tube and vortex for 10 sec. Centrifuge again. Take off the supernatant with a Pasteur pipette and dry the pellet (which is normally invisible) in a Speed-Vac-Concentrator or a vacuum desiccator.
12.	Dissolve the pellet in 25 μl of TE-buffer and store the DNA at −20°C. If possible, analyse the template DNA before use by electrophoresis on a 1% agarose gel. Take M13 virion DNA of known concentration as a reference for determining the amount of isolated template DNA by comparison of band intensities. Do not use template DNA with pronounced amounts of minichromosomes or tRNA, since these will often cause additional bands in the sequence ladder.
13.	To store clones, transfer the remaining 1 ml of each infected culture (Step 3) into a microcentrifuge tube. Centrifuge for 5 min in a microcentrifuge and pour the supernatant into a fresh tube. Do not try to transfer the supernatant quantitatively. Store this phage suspension at 4°C.

[a]*E. coli* BMH 71-18 is described in footnote [b] to *Table 1*.
[b]See footnote[b] of *Table 5*.

(iv)   Re-infection of *E. coli* with the mixed phage population at a very low multiplicity of infection to secure complete marker segregation.

The procedures are summarised in *Tables 6* and *7*.

2.4 Analysis

Individual phage clones obtained from the segregation step can now be tested for presence of the desired mutation. This is economically and most reliably carried out by direct DNA sequence analysis performed on a small number of candidate clones. The chain termination method according to F.Sanger and colleagues is especially suited for analysing DNA sequences contained in (recombinant) M13 genomes (16,17). When using this method, however, one has to synthesise a second oligonucleotide primer for the DNA sequence analysis.

Construction of Mutations in Recombinant Filamentous Phage

Table 9. DNA Sequence Analysis by the Chain Terminating Method.

1. Prepare 0.5 mM solutions of dATP, dGTP and dTTP and use to prepare the following four mixtures:

	A°	C°	G°	T°
0.5 mM dATP	1 μl	20 μl	20 μl	20 μl
0.5 mM dGTP	20 μl	20 μl	1 μl	20 μl
0.5 mM dTTP	20 μl	20 μl	20 μl	1 μl
Water	20 μl	20 μl	20 μl	20 μl

2. Prepare solutions of 2′,3′-dideoxynucleoside-5′-triphosphates: 0.1 mM of ddATP, 0.1 mM of ddCTP, 0.3 mM of ddGTP and 0.5 mM of ddTTP.

3. Mix equal volumes of A° and 0.1 mM ddATP, of C° and 0.1 mM ddCTP, of G° and 0.3 mM ddGTP and of T° and 0.5 mM ddTTP. These mixtures are henceforth called A-mix, C-mix and so on.

4. Prepare a mixture of all four deoxynucleosidetriphosphates, each at a concentration of 0.25 mM. This is called 'chase-mix'.

5. To anneal the primer and template, mix 250 ng (100 fmol) of template DNA (usually 2.5 μl of the DNA as prepared in *Table 8*), 100−200 fmol of sequencing primer (dissolved in 1−4 μl), 1 μl 500 mM NaCl, 100 mM Tris-HCl pH 7.5, 100 mM MgCl$_2$, 10 mM DTT in a microcentrifuge tube. Add water to give a final volume of 10 μl. Heat the mixture to 65°C for 5 min. Let the mixture cool down slowly to 20−30°C over a 1−2 h period. Keep on ice.

6. While waiting, prepare a denaturing polyacrylamide gel with a buffer gradient as described by Biggin *et al.* (21).

7. Prepare four tubes for each sequence reaction labelled G, A, T and C. Add 2 μl of G-mix, A-mix, T-mix and C-mix, respectively.

8. To the primer/template mixture add 1 μl [γ-^{32}P]dCTP (400 mCi/mmol; 10 μCi/μl) and 1 μl of the Klenow fragment of a DNA polymerase (1 u/μl).

9. Add 2.5 μl of this mixture to the rim of each of the four reaction tubes. Spin the drop to the bottom in a microcentrifuge to start the reaction.

10. After 15 min incubation at room temperature, add 1 μl of the 'chase-mix' in the same manner as in step 9 and incubate for another 15 min at room temperature.

11. Stop the reaction [b] by adding 45 μl of 3 M ammonium acetate and 250 μl of cold ethanol. Mix and place at −20°C for 30 min.

12. Centrifuge the tubes for 15 min at 15 000 r.p.m. in a Sorvall SS34 rotor or its equivalent at 0°C. Take off the supernatant with a drawn out Pasteur pipette.

13. Re-centrifuge for 1 min in a microcentrifuge and remove all residual traces of the supernatant with a drawn out capillary pipette.

14. Dry the pellet in a Speed-Vac-Concentrator or a vacuum desiccator, and then dissolve it in 2 μl of 80% formamide, 10 mM NaOH, 1 mM EDTA, 0.1% xylene cyanol FF, 0.1% bromophenol blue. Heat for 2 min to 95°C and apply to the denaturing polyacrylamide gel.

---

[a]1 u of the Klenow fragment of DNA polymerase I is the amount of enzyme that incorporates 10 nmol of nucleotides into an acid-precipitable fraction in 30 min. If the enzyme is at a higher concentration, dilute with 50 mM potassium phosphate pH 7.2, 50% glycerol.

[b]As an alternative way to stop the sequencing reaction (after Step 10), one can add 6 μl deionised formamide containing 0.3% xylene cyanol FF, 0.3% bromophenol blue and 20 mM EDTA. Heat 3 min to 95°C and apply 2 μl to the denaturing polyacrylamide gel.

Tables 8 and *9* summarise slightly modified versions of published sequencing procedures (17). In case of unexpectedly low marker yield, the sequencing procedure can be restricted to just one indicative reaction and more than a dozen clones can be viewed on a single gel.

After choosing a positive clone, a complete DNA sequence analysis of the entire region that was left single stranded in the gdDNA or, preferably, of the entire gene under scrutiny must be performed: since the *in vitro* DNA polymerase reaction is error prone, occasionally a clone with an additional mutation can be found (B.Kramer, unpublished).

3. ACKNOWLEDGEMENTS

Many of the procedures described in this article result from the dedicated work of my colleagues, W.Kramer, V.Drutsa, B.Kramer and M.Pflugfelder. Work in my laboratory was supported by Deutsche Forschungsgemeinschaft through Sonderforschungsbereich 74.

4. REFERENCES

1. Hutchison,C.A.,III and Edgell,M.H. (1971) *J. Virol.*, **8**, 181.
2. Zoller,M.J. and Smith,M. (1983) *Methods in Enzymology*, **100**, 468.
3. Wilkinson,A.J., Fersht,A.R., Blow,D.M., Carter,P. and Winter,G. (1984) *Nature*, **307**, 187.
4. Messing,J. (1983) *Methods in Enzymology*, **101**, 20.
5. Kramer,W., Schughart,K. and Fritz,H.-J. (1982) *Nucleic Acids Res.*, **10**, 6475.
6. Loeb,L.A. and Kunkel,Th.A. (1982) *Annu. Rev. Biochem.*, **52**, 429.
7. Wagner,R.E.,Jr. and Meselson,M. (1976) *Proc. Natl. Acad. Sci. USA*, **73**, 4135.
8. Radman,M., Wagner,R.E.,Jr., Glickman,B.W. and Meselson,M. (1980) in *Developments in Toxicology and Environmental Sciences*, Vol. **7**, *Progress in Environmental Mutagenesis*, Alacvic,H. (ed.), Elsevier North Holland, Amsterdam, pp. 121.
9. Pukkila,P.J., Peterson,J., Herman,G., Modrich,P. and Meselson,M. (1983) *Genetics*, **104**, 571.
10. Herman,G.E. and Modrich,P. (1982) *J. Biol. Chem.*, **257**, 2605.
11. Beck,E. and Zink,B. (1981) *Gene*, **16**, 35.
12. van Wezenbeek,P.M.G.F., Hulsebos,T.J.M. and Schoenmakers,J.G.G. (1980) *Gene*, **11**, 129.
13. Kramer,B., Kramer,W. and Fritz,H.-J. (1984) *Cell*, in press.
14. Lu,A.-L., Clark,S. and Modrich,P. (1983) *Proc. Natl. Acad. Sci. USA*, **80**, 4639.
15. Kramer,W., Drutsa,V., Janven,H.-W., Kramer,B., Pflugfelder,M. and Fritz,H.-J., (1984) *Nucleic Acids Res.*, **12**, 944.
16. Sanger,F., Nicklen,S. and Coulson,A.R. (1977) *Proc. Natl. Acad. Sci. USA*, **74**, 5463.
17. Walker,J.E. and Gay,N.J. (1983) *Methods in Enzymology*, **97**, 195.
18. Kornberg,A. (1980) *DNA Replication*, W.H.Freeman, San Francisco, pp. 476.
19. Gait,M.J. ed. (1984), *Oligonucleotide Synthesis: A Practical Approach*, published by IRL Press, Ltd., Oxford.
20. Maniatis,T., Fritsch,E.F. and Sambrook,J. (1981) *Molecular Cloning*, published by Cold Spring Harbor, New York.
21. Biggin,M.D., Gibson,T.J. and Hong,G.F. (1983) *Proc. Natl. Acad. Sci. USA*, **80**, 3963.

CHAPTER 9

Broad Host Range Cloning Vectors for Gram Negative Bacteria

F.CHRISTOPHER H. FRANKLIN

1. INTRODUCTION

Gram negative bacteria display a remarkable range of metabolic pathways which are of scientific, agricultural, environmental, medical and commercial importance. These include such diverse activities as the ability of *Rhizobium* to fix nitrogen (1), of *Thiobacillus* to oxidise both sulphur and iron (2), and of *Pseudomonas* to metabolise exotic organic compounds such as xylenes, naphthalenes, phenols and halogenated compounds (3). Generally, the genetic basis of these abilities is poorly understood, mainly because the organisms themselves are often not well characterised. Gene cloning technology provides a powerful tool with which to analyse and ultimately commercially exploit these activities. The most highly developed gene cloning system is based on *Escherichia coli*, its plasmids and phage. This system is perfectly adequate for the analysis of Gram negative bacteria which are closely related to *E. coli*, but for non-enteric species such as *Pseudomonas*, it has limitations. Existing evidence indicates that *E. coli* is very inefficient at expressing genes from unrelated Gram negative bacteria. If the entire toluene/xylene catabolic pathway from *P. putida* mt-2 is introduced into *E. coli*, for example, the genes are expressed at only 1−5% of the normal level in *Pseudomonas* (4). Consequently, expression studies of genes cloned from these bacteria are difficult in *E. coli*. Moreover, the phage and plasmid vectors which have been developed for *E. coli* are based on narrow host range replicons. In the case of non-enteric bacteria, this precludes the re-introduction of cloned DNA into its natural host in order to carry out expression studies. Such experiments are often necessary, especially when the activity under investigation normally occurs in a particular environment, to which the natural host bacterium is specifically adapted. Iron oxidation by *T. ferrooxidans*, for instance, will only occur in an extremely acid habitat. These constraints are not limited to the use of *E. coli* for analysis of genes from Gram negative bacteria. Similar problems exist if one's interest is in yeast or *Bacillus* genes. In all cases the answer has been the development of cloning vectors which will replicate in the organism of interest. This article outlines the development and use of plasmid vectors which are suitable for use in a wide range of Gram negative bacteria.

2. BROAD HOST RANGE PLASMIDS

Many species of bacteria have indigenous plasmids and phage, which in principle could serve as a basis for the construction of cloning vectors for the host organism. Considerable time and effort is required to develop a useful vector from a basic replicon. Therefore, it is less than desirable to construct a vector system specific to every species of interest. Fortunately, in the case of Gram negative bacteria a number of plasmids exist which have the ability to replicate in a wide range of species. These are known as broad host range plasmids and several have now been developed into cloning vectors which are useful for genomic cloning experiments in many, if not all, Gram negative bacteria.

Some examples of broad host range plasmids are presented in *Table 1*. Although they share the broad host range property other notable differences do exist. Representatives are found in several incompatibility groups, these include P-1, P-4/Q, P-6, P-9 and W. Historically, many were identified, isolated and characterised because they are of medical importance, in that they encode resistance genes to many antibiotics. For example, RP1 is a plasmid which encodes resistance to penicillin, kanamycin and tetracycline. It was first identified as the agent conferring carbenicillin resistance to a strain of *P. aeruginosa* responsible for infections in a hospital burns unit (5). However, other phenotypes are known to be associated with the plasmids such as the ability to catobalise aromatic compounds. The plasmid pJP4, for instance, encodes enzymes for 2,4-dichlorophenoxyacetic acid degradation (6). Some of these plasmids, notably RK2 (also known as RP1, RP4, R68 and R18) (5), are conjugative, whilst others, for example RSF1010 (7), are non-transmissable. Considerable DNA coding capacity is required to specify conjugal transfer, which to a great extent explains the comparatively large size of RK2 (60 kb) and its relatives. As one might anticipate, the non-conjugative plasmids have a reasonably low molecular weight RSF1010, for example, is 8.9 kb in size. Logically, it might be imagined that these small plasmids would provide the most suitable basis for the construction of very small cloning vectors but, as will become apparent, this is not entirely the case. Although the most extensive

Table 1. Some Broad Host Range Plasmids.

Plasmid	Incompatibility group	Genome size (kb)	Phenotypic characteristics[a]
RK2, RP1, RP4, R68, R18	P-1	60	*Ap, Km, Tc, Tra*$^+$
RSF1010, R300B, R1162	P-4/Q	8.9	*Su, Sm, Tra*$^-$, *Mob*$^+$
Sa	W	29.6	*Km, Cm, Sp, Su, Tra*$^+$
pVS1	Unclassified	30	*Hg, Su*; fails to replicate in *E. coli*
RMS149	P-6	54	*Cb, Gm, Sm, Su*
pJP4	P-1	52	*Hg*, growth on 2,4-D
pWWO	P-9	117	Growth on toluene/xylene

[a]Abbreviations used for resistance and phenotypic markers: *Ap*, ampicillin; *Km*, kanamycin; *Tc*, tetracycline; *Su*, sulphonamide; *Sm*, streptomycin; *Cm*, chloramphenicol; *Sp*, spectinomycin; *Hg*; mercuric ions; *Gm*, gentamycin; *Tra*$^+$, transfer proficient; *Mob*$^+$, ability to be co-transferred by a conjugative plasmid; 2,4-D, utilisation of 2,4-dichlorophenoxyacetic acid.

range of vectors has been developed from the incQ plasmid RSF1010 (7), useful vectors derived from the considerable larger Sa (29.6 kb) and RK2 (60 kb) plasmids are also available (5,14,15). The development and methodology for using the various vector systems are very similar. For this reason only one of them, namely RSF1010, will be described in detail. The others will be mentioned only in terms of what plasmids are available and what special features they offer.

2.1 Development of incQ/W Plasmids as Broad Host Range Cloning Vectors

The incQ plasmids RSF1010, R300B and R1162 are three, almost identical, 8.9 kb non-conjugative, multicopy replicons which specify resistance to streptomycin (*Sm*) and sulphonamides (*Su*). A detailed physical and functional map of RSF1010 (*Figure 1*) has been constructed. The features which have been mapped include restriction endonuclease recognition sites, RNA polymerase binding sites, genes for *Sm* and *Su* resistance, plasmid mobilisation (*Mob*), three replication proteins (*Rep* A, B and C) and the origins of vegetative (*Ori*) and transfer (*Nic*) replications (9,10). Restriction endonuclease mapping of the plasmid indicated that some modifications to the basic replicon were necessary before it would be generally useful as a vector system. In common with other broad host range plasmids,

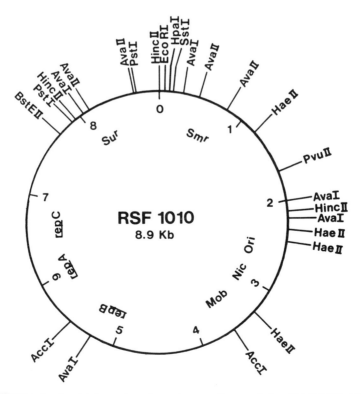

Figure 1. Functional and restriction endonuclease cleavage site map of plasmid RSF1010. Abbreviations: Su^r, sulphonamide resistance; *Sm*^R, streptomycin resistance; *rep*A, B and C, replication genes; Mob, ability to mobilise; Nic, relaxation nick site; Ori, origin of replication.

RSF1010 has few unique restriction endonuclease cleavage sites suitable for gene cloning experiments. It is therefore desirable to incorporate more single sites into the plasmid. Four potentially useful cleavage sites are already present, namely *Bst*-EII, *Sst*I, *Eco*RI and *Pst*I (two adjacent sites). However, before the latter three may be used, another selective marker must be introduced into the RSF1010 genome. The problem occurs because *Su* and *Sm* are transcribed from the same promoter (9,11). Insertion of a DNA fragment between the *Pst*I sites which straddle the promoter sequences inactivates both *Su* and *Sm* resistance determinants. Cloning at the *Sst*I site destroys *Sm* resistance. Although *Eco*RI is not located within the coding sequences of the *Sm* gene, resistance is lost if a DNA segment is inserted at this site, unless it provides a new promoter for transcription of the gene. Thus when DNA is inserted into the *Eco*RI or *Sst*I sites, although *Su* resistance remains, it is a less than ideal selective marker. By incorporating DNA fragments encoding antibiotic resistance genes into the RSF1010 replicon, vector plasmids suitable for cloning with a range of different restriction endonucleases have now been constructed.

The derivation of three of these, namely pKT210, pKT231 and pKT230, is presented in *Figure 2*. A range of broad host range cloning vectors is given in *Table 2*. Plasmid pKT210 was constructed by cloning a 3.6 kb DNA fragment encoding chloramphenicol (*Cm*) resistance from plasmid Sa between the *Pst*I sites of RSF1010. The vector may be used for cloning with *Eco*RI, *Sst*I and *Hind*III. Recombinant molecules containing *Eco*RI or *Sst*I generated fragments are detected by insertional inactivation of *Sm* resistance. Plasmids pKT230 and pKT231 are quite similar; pKT230 is a double replicon constructed by ligating *Pst*I linearised pACYC177 (12) with *Pst*I digested RSF1010. It encodes *Sm* and kanamycin (*Km*) resistances and has cloning sites for fragments generated by *Hind*III, *Xho*I, *Xma*I, *Bst*EII, *Bam*HI, *Eco*RI or *Sst*I. The vector pKT231 also encodes *Sm* and *Km* resistance on a 4.6 kb *Pst*I fragment derived from the R6-5 mini-plasmid pKT105 and incorporated into RSF1010. It is possible to clone *Eco*RI, *Sst*I, *Hind*III, *Xho*I, *Xma*I and *Cla*I fragments into this plasmid. All the vectors based on the RSF1010 replicon have a copy number of 15 − 20 per genome equivalent.

Vectors have been constructed in which a tetracycline (*Tc*) resistance gene has been incorporated in the RSF1010 genome as a selective marker. Although plasmids pKT212 and pKT214 are included in *Table 1*, it must be noted that they are of limited value for use in *Pesudomonas* and possibly other Gram negative bacteria. They appear to be unstable, particularly in bacterial cultures grown in the absence of antibiotic selection, as it is thought the location of the *Tc* resistance protein in the cell membrane has an adverse effect on the host viability (13).

It will be noticed that all the vectors based on RSF1010 are large compared with the narrow host range *E. coli* vectors in routine use. During the early development of the RSF1010 vectors, numerous attempts were made to reduce the size of the plasmid replicon (9,10). These were not successful. Recent studies (10) have explained the problem, which lies in the complexity of the plasmid replication functions. It appears that at least three replication genes are required, *rep*A, *rep*B and *rep*C and these are located on the plasmid genome several kb distant

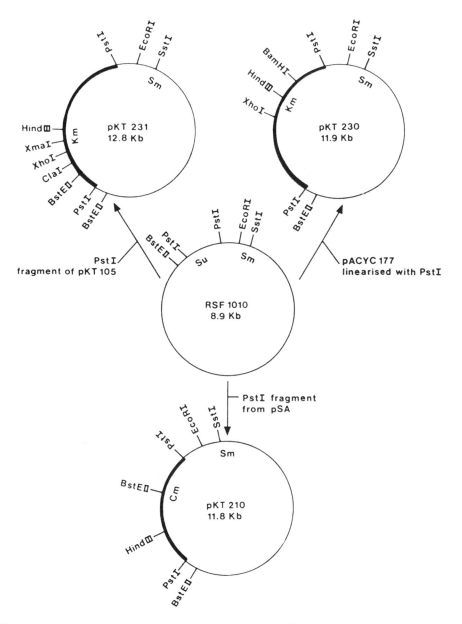

Figure 2. Broad host range cloning vectors derived from RSF1010.

from the origin (*ori*) of replication (*Figure 1*). Therefore, little of the plasmid genome can be considered dispensible. This makes the goal of constructing vectors of comparable size to the *E. coli* plasmids extremely difficult.

Plasmid vectors based on RSF1010 have now been introduced into an extensive range of Gram negative bacteria (*Table 3*).

Table 2. Broad Host Range Cloning Vectors.

Vector	Replicon(s)	Size (kb)	Cloning sites	Marker for selection[a]	Insertional inactivation	Special features[b]	Reference
pKT210	RSF1010	11.8	SstI	Cm	Sm	Mob[+]	9
			HindIII	Cm,Sm	−		
			EcoRI,HpaI	Cm	Sm		
pKT230	RSF1010/ pACYC177	11.9	SstI,EcoRI	Km	Sm	Mob[+]	9
			HindIII,XhoI,XmaI	Sm	Km		
			BstEII,BamHI	Km,Sm	−		
pKT231	RSF1010	12.8	SstI,EcoRI	Km	Sm	Mob[+]	9
			HindIII,XhoI,XmaI, ClaI	Sm	Km		
pKT212/214	RSF1010	15.7	BglII,XbaI	Sm,Cm	Tc	Mob[+]	13
			BamHI	Sm,Cm,Tc	−		
			SstI,EcoRI	Cm,Tc	Sm		
pKT262	pKT230	11.7	Deletion derivative of pKT230			Mob[−]	9
pKT263	pKT231	12.8	Deletion derivative of pKT231			Mob[−]	9
pKT240	RSF1010	12.5	HindIII,XhoI,XmaI, ClaI	Ap	Km	Promoter probe vector Mob[+]	32
			SstI,EcoRI	Ap,Km	−		
pMMB33/34	RSF1010	13.75	BamHI,EcoRI, SstI	Km		Cosmid Mob[+]	30
pMMB22	pKT240	12.7	EcoRI	Ap		tac pro-motor expression vector[c]	32
pMMB24	pKT240	12.7	HindIII	Ap	−	tac pro-moter expression vector[c]	32
pGV1106	Sa	8.4	EcoRI	Km	Sp		15
			BamHI,PstI,BglII, SacII	Km,Sp	−		
pGV1122	Sa	10.8	HindIII,BamHI, SalI	Sp	Tc	Ampli-fiable by Cm	15
			PstI	Sp,Tc	−	Mob[+]	
pSa4	Sa	9.4	KpnI	Sp,Cm	Km		14
			BamHI,SstI	Sp,Cm,Km	−		
pRK290	RK2	20	EcoRI,BglII	Tc	−	Low copy no. Mob[+]	20
pRK2501	RK2	11.1	XhoI,HindIII	Tc	Km	Low copy no. Mob[+]	20
			SalI	Km	Tc		
			EcoRI,BglII	Km,Tc	−		

Table 2 continued.

Vector	Replicon(s)	Size (kb)	Cloning sites	Marker for selection[a]	Insertional inactivation	Special features[b]	Reference
pRP301	RP4	54.7	BamHI,EcoRI, HindIII	Ap,Tc	–	Low copy no. Tra⁺	20
pLAFR1	pRK290	21.6	EcoRI	Tc	–	Cosmid Mob⁺	28
pME290	pVS1	6.8	HindIII,XhoI, XmaI PstI	Cb Km	Km Cb	Will not replicate in E. coli	20

[a,b]Abbreviations are the same as indicated in the footnote to *Table 1*.
[c]*tac* is a promoter fusion between the −35 region of the E. coli *trp* promoter and the −10 region of the E. coli *lac* UV5 promoter (32).

2.2 Vectors Based on the IncW Sa Plasmid

A versatile range of cloning vectors has been constructed from pSa by the groups of Kado (14) and Schell (15). The plasmid pSa has a molecular weight of 29.6 kb and belongs to incompatibility group W. The vectors derived from it are presented in *Table 2*. The initial step in their development was the isolation of a 'mini-Sa' by deletion of the conjugal transfer genes from the plasmid, following which further modification was carried out to introduce selectable markers and create suitable cloning sites. In contrast to RSF1010, isolation of 'mini-Sa' plasmids has proved straightforward. The plasmid pGV1106 (15), for example, which encodes *Km* and *Sp* was isolated by recircularising an 8.4 kb *BglII* fragment of pSa.

The pSa-derived vectors provide a useful alternative to the RSF1010 vectors. Furthermore, as they belong to different incompatibility groups, they can be stably maintained in the same bacterial cell. This creates the possibility of performing genetic analyses, such as complementation studies between cloned sequences in a wide range of Gram negative hosts.

3. PRACTICAL USE OF BROAD HOST RANGE VECTORS

In this section the basic methodology for using the vectors will be described.

3.1 Isolation of Vector DNA

Most methods in common use are satisfactory for isolation of vector DNA, particularly if the plasmid is in an E. coli background. If DNA is to be used in gene cloning experiments, purification by banding in a CsCl-ethidium bromide gradient is advised. Reproducibly good yields may be obtained by following the 'cleared lysate' method described in reference 16. However, there are two points to note. First, with a few exceptions the vectors do not undergo additional replication in comparison with the chromosomal DNA when chloramphenicol is added to the culture medium. Second, Gram negative bacteria such as *Pseudomonas* lyse more rapidly than E. coli. It is, therefore, necessary to reduce the incubation time following the addition of detergent (Brij58 or Triton X-100) to 1−2 min.

Table 3. Host Range of RSF1010 Derived Vectors[a,]

Escherichia coli
Pseudomonas aeruginosa
Pseudomonas putida
Agrobacterium tumefaciens
Azotobacter vinelandii
Alcaligenes eutrophus
Alcaligenes faecalis
Acetobacter xylinum
Methylophilus methylotrophus
Rhodopseudomonas spheroides
Rhizobium meliloti
Rhizobium leguminosarum
Caulobacter crescentus
Yersinia enterocolitica
Klebsiella aerogenes
Serratia marcescens
Erwinia coratovora
Xanthomonas compestris
Gluconobacter sp.

[a]Compiled from references 9 and 13.

If this precaution is not observed, soft pellets are obtained when the 'cleared lysate' is centrifuged. This results in considerable loss of plasmid DNA and makes it technically difficult to remove the lysate from the pellet of cell debris.

'Mini-prep' methods, particularly those of Holmes and Quigley (17) and Birnboim and Doly (18), are very useful for checking plasmid DNA from transformants. Again, difficulties might be encountered when isolating the plasmids from bacteria other than *E. coli* because of problems with cell lysis. But, generally speaking, the Birnboim and Doly method works well with most bacteria tested to date.

3.2 Cloning DNA Fragments Using Broad Host Range Vectors

The methodology of DNA cloning using plasmid vectors is well documented (for example, see reference 19). No special considerations are required when using broad host range vectors. Except that, as they are generally larger than the narrow host range *E. coli* vectors, it is necessary to use a higher concentration of linearised vector DNA in the ligation mix to achieve the desired molar concentration of 'sticky' or 'flush' ends.

3.3 Transformation of Non-Enteric Gram Negative Bacteria with Broad Host Range Vector DNA

Although efficient methods for the transformation of *E. coli* with plasmid DNA have been developed (see for example reference 9), there are few documented examples of transformation of non-enteric Gram negative bacteria. A number of bacteria such as *P. stutzeri* and *Thiobacillus* sp. can undergo transformation by linear double-stranded DNA, without prior chemical treatment of the bacterial

Table 4. Transformation Frequencies of Various Bacteria with RSF1010 DNA.

Host bacteria[a]	Relevant phenotype[b]	Transformation frequency (per µg RSF1010 DNA)
E. coli SK1592	hsdR hsdM$^+$	3.6 x 10^6
P. aeruginosa PAO1162	rmo mod$^+$	6.7 x 10^4
P. putida met-2 KT2440	rmo mod$^+$	3 x 10^3
P. putida mt-2	rmo$^+$ mod$^+$	<3.0 x 10^1 (increases to 8 x 10^3 if host modified DNA is used)

[a]See reference 9.
[b]*hsd*R, host specific restriction; *hsd*M$^+$, host specific modification; *rmo*, host specific restriction; *mod*$^+$, host specific modification.

cells (20). For plasmid transformation a major factor in the efficient transformation of *E. coli* is the availability of restriction defective (*hsd*R) recipients. Such mutants of non-enteric bacteria are not usually available, and consequently it is extremely difficult, if not impossible, to transform them with heterologous DNA. However, in the case of *P. aeruginosa* and *P. putida*, bacteria with highly efficient restriction systems (9,21), restriction-less mutants (*rmo*) have been isolated (9). *Table 4* illustrates the efficiency with which these *Pseudonomas* mutants may be transformed by plasmid DNA. For comparison, the figures for transformation of an *E. coli hsd* strain SK1592 and *P. putida* mt-2 *rmo*$^+$ are included. It may be seen that the efficiencies for the *P. aeruginosa* PaO1162 *rmo* and *P. putida* 2440 *rmo* are considerably lower than that for *E. coli*. In the case of *P. putida* mt-2 *rmo*$^+$ transformed with heterologous DNA they are lower still, but it is interesting to note that if the host modified plasmid DNA is used the frequency is similar to that obtained with the restriction defective mutant.

The problem of poor transformation efficiency makes it, for most practical purposes, unrealistic to attempt direct transformation of non-enteric bacteria with DNA from a ligation. The most satisfactory approach is to make use of *E. coli* as an intermediate host for transformation of the ligation mix and screening for recombinant plasmids. Once identified, the recombinant plasmid DNA of interest can be purified and then used to transform the desired host. The isolation of the recombinant DNA in this way acts as an amplification step, thereby increasing the chances of successfully transforming the host of choice. The transformation method presented here is based on that of Kushner (22) and has been optimised to achieve reasonably efficient transformation of *Pseudomonas* type bacteria with purified DNA (9).

(i)   Grow the bacteria in L-broth (23) to an OD$_{650nm}$ of 0.3−0.4.
(ii) Harvest cells by centrifugation at 7000 r.p.m. for 5 min at 0°C. Discard the supernatant.
(iii) Gently resuspend the cells in equal volumes of ice cold transformation buffer (10 mM MOPS, pH 7.0, 10 mM RbCl, 100 mM MgCl$_2$) and centrifuge as (ii).

(iv)   Gently resuspend the cells in an equal volume of ice cold transformation buffer and incubate for 30 min at 0°C.

(v)   Centrifuge the cells, discard the supernatant and resuspend the cells in 1/10 volume ice cold transformation buffer.

(vi)   Aliquot the resuspended cells into 0.2 ml portions and add DNA (up to 1 µg in a maximum volume of 25 µl). Incubate for 45 min at 0°C.

(vii) Heat shock the cells at 42°C for 2 min.

(viii) Add 2 ml L-broth to the cells and incubate at 30°C for 90 min.

(iv)   Plate out 0.1 ml aliquots on appropriate selective media.

It may be noted that the transformation buffer contains no $CaCl_2$. Although this is commonly employed for transformation of *E. coli* and has been used in transformation of *Pseudomonas* sp., it appears to have a killing effect on *Pseudomonas*, reducing the transformation frequency ~10-fold (M.Bagdasarian and F.C.H.Franklin, unpublished results).

3.4 Mobilisation of Broad Host Range Vectors as a Method of Transfer to Recipient Bacteria

At present it is not possible to transform many species of non-enteric bacteria. Probably, as various species are studied in greater detail, the appropriate methods will be developed. Until such a time, in the absence of an alternative, use of recombinant DNA techniques for genetic analysis of these bacteria would be limited. Fortunately, there is an alternative to transformation as a means of introducing recombinant plasmids into recipient bacteria, namely plasmid mobilisation. The method also has advantages for the non-enteric bacteria that can be transformed, as it avoids the requirement of purifying DNA which is generally a pre-requisite for successful transformation of *Pseudomonas* and related bacteria. This can be a considerable problem if it is necessary to introduce a large number of recombinants into a recipient, as it is extremely time consuming to isolate plasmid DNA from perhaps several hundred clones. Although it is possible to prepare plasmid DNA from 'pooled' clones and use this for transformation, it is generally quicker and more straightforward to use plamid mobilisation. Many of the plasmid vectors described in this article, whilst not self-transmissible, can be transferred to a wide range of recipients by mobilisation. This property depends on the presence on the vector replicon of the *mob*, *ori*T (origin of transfer) and *nic* (nick induced by the relaxation complex) determinants. These regions have been mapped for plasmids RSF1010 (9) and RK2 (5). The other component of the system is the presence in the donor cell of a helper plasmid which provides the necessary transfer (*tra*) genes required for mobilisation of the vector to a recipient strain. Various conjugative plasmids fill the role of 'helper' (see *Table 5*) with high efficiency. RP4, for example, will mobilise RSF1010 derivatives such as pKT230 with a frequency transfer of 7.5×10^{-1} per donor cell (9). One finding that is both surprising and technically useful is the ability of incIα plasmids such as pLG221 to mobilise the vectors to a wide range of recipients (24; Franklin and Boulnois, unpublished). The incIα plasmids have a host range restricted to *E. coli* and its close relatives. Consequently, when they are used to mobilise plasmids to species

Table 5. Some Helper Plasmids for Vector Mobilisation.

Plasmid	Replicon	Incompatibility group	Selective[a] markers	Reference
pLG221	ColI-b drd-1	Iα	Km	24
pLG223	ColI-b drd-1	Iα	Tc	24
R64 drd-11	R64	Iα	Sm	9
RK2/RP4	–	P-1	Tc, Ap, Km	5
pSa322	Sa/pBR322	W	Ap	14

[a]Abbreviations are indicated in the footnote to *Table 1*.

like *Pseudomonas* they cannot themselves become established in the recipient. Thus, transfer of the vector or recombinant plasmid can be achieved without the inconvenience of co-transfer of the helper plasmid. Two methods are presented here, both are similar. The former is most suitable for transfer of single or a few clones, whereas the latter is adapted for experiments which demand the mobilisation of large numbers of plasmid clones, for instance, screening a cosmid library.

3.4.1 *Mobilisation of single clones*

(i) Grow the donor (strain containing the plasmid to be mobilised and the 'helper' plasmid) and the recipient in separate 5 ml cultures in L-broth at 30°C overnight. The medium in which the donor bacteria are grown should contain appropriate antibiotics, resistance to which is determined by the plasmids.

(ii) Place a sterile 10 mm diameter nitrocellulose filter (e.g., Millipore HAWPO1300 pore size 0.45 μm or its equivalent from other suppliers) on an L-agar plate.

(iii) Pipette 10 μl each of donor and recipient onto the filter. Incubate 6 – 15 h at 30/37°C.

(iv) Transfer the filter to a test tube containing 1.0 ml of 0.9% NaCl. Vortex to resuspend the cells.

(v) Dilute cell suspensions and plate on media appropriate for selecting the recipient strain carrying the plasmid.

3.4.2 *Mobilisation of Large Numbers of Clones*

(i) Inoculate individual colonies of donors into 96 well micro-titre plates which contain 0.1 ml aliquots of L-broth and appropriate antibiotic. Incubate overnight at 37°C.

(ii) Grow the recipient bacteria in 5 ml L-broth overnight at 30/37°C.

(iii) Spread 0.1 ml of recipient culture on the surface of an L-agar plate. Two plates are required per micro-titre plate of donors.

(iv) Transfer small aliquots (5 – 10 μl) of the donor strains to the surface of the plate spread with recipient bacteria. This is most effectively achieved using a sterile 48 tooth comb, the teeth of which are arranged in a block of 6 x 8 such that they fit into half the wells of the micro-titre dish. This allows transfer of 48 clones in a single operation. Incubate the plates 6 – 12 h at 30/37°C.

(v)   Replica plate (23) onto media appropriate for selecting the recipient strain containing the mobilised plasmid.

4. SPECIAL PURPOSE BROAD HOST RANGE CLONING VECTORS

Many *E. coli* vectors now available incorporate particular features which make them suited for a specific purpose. Typical examples are expression vectors (19), plasmids for cloning sequences such as promoters and terminators (25), those which allow direct detection or selection of recombinants (26), low copy number vectors and plasmids (cosmids) suitable for cloning long stretches of DNA for the construction of genomic libraries (27). Comparable broad host range vectors are now becoming available. As yet, they are not as sophisticated as their *E. coli* vector counterparts, but they can often offer advantages for analysis of genes from unusual bacteria.

4.1 **Broad Host Range Cosmids**

The λ *in vitro* packaging procedure allows one to use cosmids (plasmids containing the λ *cos* site) to clone DNA fragments of 30−40 kb into an *E. coli* host. Potentially this is an extremely useful method for the genetic analysis of the more exotic activities of Gram negative bacteria. A two stage procedure has now been developed. It is based on an RSF1010 cosmid used in conjunction with the mobilisation system described in Section 3.4.2.

Although a number of broad host range cosmids have been constructed (28,29), with the exceptions of pMMB33 and pMMB34 (30) all have limitations. Some have been shown to be unstable; others lack a single *Bam*HI or *Bgl*II restriction site necessary for cloning random DNA fragments generated by partial digestion with *Sau*3A or *Mbo*I. Furthermore, they all generate polycosmids during the ligation step unless the cloning site is dephosphorylated.

Cosmids pMMB33 (*Figure 3*) and pMMB34 are 13.75 kb vectors based on an RSF1010 replicon. They differ only in respect of the orientation of the *Bst*EII/*Hpa*I fragment, on which are located the *Km* resistance gene and the λ *cos* site. An important step in the construction of these plasmids was the conversion of the *Hpa*I site in RSF1010 (located between *Eco*RI and *Sst*I) to a *Bam*HI site, using synthetic linkers. These vectors have been successfully used to construct genomic libraries of several Gram negative bacteria including a number of *Pseudomonas* species and a *Thiobacillus* species (30).

4.1.1 *Cloning in pMMB33 and pMMB34*

An outline of the cloning strategy is presented in *Figure 3*.

(i)   Divide 5 µg of vector DNA (the quantity can be varied as required) into two aliquots. Digest with *Hpa*I and the other with *Sma*I.

(ii)   Terminate the reaction by incubating at 70°C for 10 min.

(iii)   Run a small aliquot from each digest on an agarose gel, to check that the digests are complete.

(iv)   Digest both samples to completion with *Bam*HI.

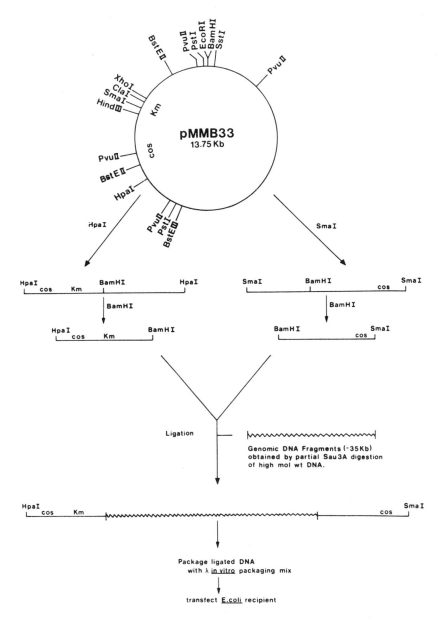

Figure 3. Cosmid pMMB33 and outline of cosmid cloning scheme.

(v)   Partially digest high mol. wt. (~100 kb) genomic DNA with *Sau*3A in conditions which generate a majority of 30−40 kb sized fragments. At this stage the chromosomal fragments can be size selected on a sucrose or NaCl gradient. However, in practice this is not usually necessary for construction of genomic libraries from bacteria, as they have relatively small genomes and consequently the required number of recombinants (~500) can be achieved without size selection on a gradient.

(vi)   Ligate 1 µg of vector 'arms' with 5 µg *Sau*3A digested chromosomal DNA in 25 µl ligation buffer containing 5 mM MgCl₂ and 5 mM ATP (30).

(vii)   Package 5 µl aliquots of the ligated DNA with a λ *in vitro* packaging mix as described by Maniatis *et al.* (19).

(viii) Grow *E. coli* DH1 (19) or CF503 (30) to late exponential phase in L-broth supplemented with 0.4% maltose. Mix 500 µl of this culture with 250 µl of *in vitro* packaged cosmids and incubate at 37°C for 1 h.

(ix)   Plate the bacteria on L-agar supplemented with *Km* (50 µg/ml) for selection.

(x)   To store the clones, transfer single colonies to 96 well micro-titre plates containing 200 µl Hogness freezing medium (L-broth supplemented with 8.8% glycerol, 3 mM trisodium acetate, 55 mM K₂HPO₄, 26 mM KH₂PO₄, 1 mM MgSO₄, 15 mM (NH₄)₂SO₄). Incubate 6 h at 37°C and then store at −70°C.

If it is necessary to screen for clones of interest in a genetic background other than *E. coli*, it is useful to transfect the packaged cosmids directly into an *E. coli* strain containing a suitable mobilising plasmid (e.g., CF503). The entire library can then be conveniently mobilised into the recipient of choice, as described in Section 3.4.2.

4.2 Broad Host Range Regulatable Expression Vectors

Once a particular gene has been cloned, it is often desirable to manipulate its expression to obtain high levels of gene product. The objective here might be to isolate the gene product in sufficient quantities for fundamental research or perhaps commercial exploitation. Numerous expression vectors derived from *E. coli* plasmids are available. They incorporate well characterised, strong promoters such as *lac* UV5, *trp* and λP$_L$ (19). An important feature of all these promoters is that their transcriptional activity is regulatable by the presence of the appropriate repressor molecule. This allows expression to be switched off until required, thereby avoiding the deleterious effects on cell growth sometimes encountered when a cloned gene is expressed at high levels continuously throughout the growth phase.

Two expression vectors based on the broad host range plasmid RSF1010 have recently been constructed (*Figure 4*). The plasmids pMMB22 and pMMB24 contain a *tac* promoter; this is a fusion between the −35 region of *E. coli trp* and the −10 region of *lac* UV5 (31,32). This promoter is regulated by the *lac* repressor. A fragment encoding the *lac*IQ gene was incorporated into the plasmids during construction, with the result that transcriptional activity of the *tac* promoters on both pMMB22 and pMMB24 is regulatable. In the case of pMMB22 the *tac* promoter was introduced between the *Eco*RI and *Hind*III sites on pKT240 (32), such that transcription proceeds through the *Eco*RI site. In pMMB24 the *tac* promoter is in the reverse orientation, with the result that DNA fragments cloned at the *Hind*III site are transcribed off *tac*.

A major question which arose during the construction of the plasmids was whether or not the *tac* promoter would function and would be regulated in species other than *E. coli*. Although it is known that genes from non-enteric Gram negative

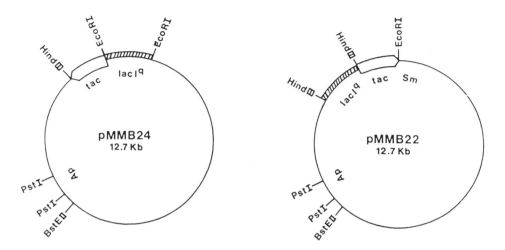

Figure 4. Tac promoter expression vectors pMMB22 and pMMB24.

Table 6. Expression Levels of Catechol 2,3-Oxygenase Obtained Using the RSF1010 Derived Expression Vector pMMB22.

Plasmid[a]	Host[b] strain	Promoter[c]	Inducer[d]	C23O activity (mU/min/μg protein)
pWWO-161	P	meta	–	170
pWWO-161	P	meta	m-MBA	3 040
pWWO-161	E	meta	–	50
pWWO-161	E	meta	m-MBA	240
pMMB25	E	tac	–	1040
pMMB25	E	tac	IPTG	20 350
pMMB25	P	tac	–	500
pMMB25	P	tac	IPTG	20 500

[a]pWWO-161, Tn401 insertion derivative of TOL plasmid pWWO, see reference 4; pMMB25, pMMB22 with the pWWO C23O gene inserted at the vector *EcoRI* site.
[b]P, *P. putida* mt-2 KT2440; E, *E. coli* SK1592.
[c]*meta*, meta-pathway promoter from pWWO. *tac*, as defined in footnote [c] of *Table 2*.
[d]m-MBA, meta-methyl benzyl alcohol 5 mM. ITPG, isopropylthiogalactoside, 2 mM for *E. coli*, 5 mM for *P. putida*.

bacteria do not express well in *E. coli*, little is known about the reverse situation. Recently, it was reported that the *E. coli* trp operon is expressed efficiently by *P. aeruginosa* (33), indicating that perhaps there is no expression barrier for genes transferred in this direction. To investigate if this was the case for genes transcribed from the *tac* promoter in the expression vectors, a fragment encoding the enzyme catechol 2,3-oxygenase (C23O) (4) was cloned into the *EcoRI* site of pMMB22 to create pMMB25 (32). The fragment encodes the C23O structural gene and Shine Dalgarno sequence (34) but no promoter. Expression of C23O was determined by enzyme assay of cell free extracts prepared from *E. coli* and *P. putida*. The results are presented in *Table 6*. It may be seen that similar levels of C23O activity are obtained in *E. coli* and *P. putida*. Furthermore, in both cases expression is subject to repression by the *lac* repressor and is induced in the presence of IPTG. This demonstrates the potential value of pMMB22 and pMMB24 for obtaining high levels of regulated expression of cloned genes in non-enteric Gram negative bacteria.

Practically, there appears to be only one significant difference between use of the plasmids in *E. coli* and *P. putida*. This is the amount of IPTG required to obtain maximal levels of induction, in the case of *E. coli* it is 2 mM but for *P. putida* a slightly higher concentration, 5 mM, is required. It is conceivable that this figure will differ in other bacteria and should therefore be optimised when using the plasmids in different organisms.

4.3 Low Copy Number Broad Host Range Vectors

In order to clone certain genes, such as those encoding membrane proteins, it is necessary to use low copy number vectors, as greatly enhanced expression of this type of gene is deleterious or even lethal for the host cell. Several low copy number vectors are available (*Table 2*). Plasmids pRK290, pRK2501 and pRP301 are derived from the apparently identical IncP-1 conjugative plasmids RK2 and RP4. The vectors have copy numbers of about 2 − 8 per chromosome equivalent (13) and retain the broad host range of the parental plasmids. No quantitative data is available regarding their stability in hosts other than *E. coli* (13).

A range of vectors (e.g., pME290) has been constructed from the *P. aeruginosa* plasmid pVS1 (35). These have a copy number of about 8 (20) and in the case of pME290 is conveniently small (6.8 kb) compared with many other broad host range vectors. A further interesting feature and point to take into account for practical use of pVS1 and its derivatives is that whilst they replicate in *Pseudomonas sp* and related bacteria, they do not in *E. coli* (20).

4.4 An RSF1010 Derived 'Promoter Probe' Vector

The construction of plasmids which can be specifically used to identify DNA signal sequences such as promoter regions and terminators has proved to be of considerable value for the analysis of prokaryotic gene regulation (25,26). Vectors designed to identify promoters in *E. coli* are based on well characterised genes which specify a gene product which can be easily identified and assayed, for example galactokinase (25) and galactosidase (36). A DNA segment that provides promoter function will cause these genes to be expressed when cloned upstream of them.

A broad host range vector has recently been described that can be used in a similar way to probe for promoters (32). The plasmid pKT240 (*Figure 5*) was constructed by replacing the 1.3 kb *Bst*EII-*Eco*RI fragment of RSF1010 with a 3.9 kb *Bst*EII-*Eco*RI segment encoding *Ap* and *Km* resistance from pHSG415 (37). The replacement of the RSF1010 sequences results in the deletion of the *Su/Sm* resistance promoter and coding sequences of the *Su* resistance structure gene. Consequently, the plasmid no longer confers resistance to *Sm*. However, incorporation of DNA fragments encoding promoter sequences into the *Eco*RI site of pKT240 restores the *Sm* resistance of the plasmid (32, Franklin, unpublished

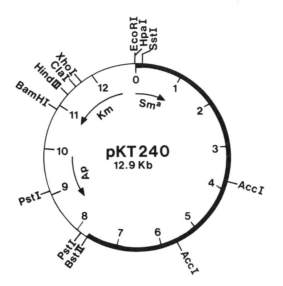

Figure 5. Promoter probe vector pKT240. [a]Streptomycin resistance is dependent on the insertion of promoter sequences at the *Eco*RI site.

data), providing an easily scored phenotype to detect promoter activity.

The plasmid can be used to screen for specific fragments which encode putative promoters, or for 'shot-gun' cloning of promoter sequences in the following way.

(i)   Digest 1 μg of pKT240 DNA to completion with *Eco*RI in 50 μl of 50 mM Tris-HCl pH 7.5, 10 mM MgCl$_2$, 100 mM NaCl.

(ii) Inactivate the restriction endonuclease by heating to 70°C for 5 min. Precipitate the DNA by adding 3 M sodium acetate (such that the final salt concentration is 0.3 M), followed by one volume of iso-propanol that has been pre-chilled to −20°C. Place at −70°C for 20 min. Centrifuge the precipitated DNA in a microcentrifuge. Carefully remove the supernatant, wash the DNA pellet with 70% ethanol and allow the pellet to dry in a vacuum desiccator.

(iii) Resuspend the precipitated DNA in 50 μl phosphatase buffer (50 mM Tris-HCl pH 9.0, 1 mM MgCl$_2$, 100 μM ZnCl$_2$, 1 mM spermidine). Incubate at 37°C for 30 min with two units of calf intestinal phosphatase to remove the 5′ phosphoryl groups and hence prevent vector recirculation during the ligation step (19). Terminate the reaction by extracting twice with an equal volume of phenol. Remove the aqueous phase and extract five times with ether to remove residual phenol. Precipitate the DNA as described in step (ii). Resuspend in 20 μl ligation buffer (50 mM Tris-HCl pH 7.5, 10 mM MgCl$_2$, 2 mM ATP, 5 mM DTT).

(iv) Digest 5 μg of the DNA to be screened with *Eco*RI as outlined in step (i), using the procedures described in Steps (ii) and (iii). Terminate the digestion, precipitate the DNA and resuspend in 80 μl ligation buffer. Mix with the vector DNA, add 0.25 unit of T4 DNA ligase and ligate overnight at 15°C.

(v)   Transform *E. coli* with the ligated DNA using a protocol such as that described in Chapter 6. Two options are open at this stage. *E. coli* can be transformed with the intention of screening promoter activity in this host. Alternatively, bearing in mind the apparent poor expression of genes from unrelated bacteria in *E. coli*, a recipient can be chosen which harbours a mobilising plasmid as outlined in Section 3.4.1. This permits transfer to the natural genetic background for the screening procedure.

(vi) In the former case, plate transformed cells onto L-agar plates containing *Sm* in a range of concentrations from 20 to 200 µg/ml. In the latter, the recombinant plasmids are mobilised to the host from where the putative promoter sequences originate. If a specific fragment is being tested, the mobilisation can be carried out after colonies containing the correct recombinant have been identified. For shot-gun experiments, mobilisation can be performed prior to plating the transformed cells. After allowing a period of time for expression (60 – 90 min) mix the transformed cells (1 ml) with 0.2 ml of a late log phase culture of the recipient bacteria. Plate 25 µl aliquots on filters and proceed as described in Section 3.4. The cells are then plated on L-agar containing 20 – 200 µg/ml of streptomycin and a suitable counter selection for the donor and incubated overnight.

(vii) Without the insertion of a promoter sequence, the vector pKT240 specifies resistance to 20 µg/ml of streptomycin in *E. coli* and 50 µg/ml of streptomycin in *P. putida*. Pick colonies which are growing at higher concentrations of streptomycin and restreak to purify and check the resistance.

(viii) Plasmid DNA can be isolated and the inserted fragment characterised. Accurate estimation of promoter activity can be determined by assay of the streptomycin phosphotransferase activity (32).

4.5 Containment Vectors Derived from RSF1010

Certain cloning experiments are subject to recombinant DNA regulations and must be performed in certified host:vector systems. One of the key regulations is that the vector involved be non-transmissible and not mobilised by co-transfer with a conjugative plasmid. A number of RSF1010 derived plasmids which meet this requirement have been constructed (9). To achieve this, an RSF1010 derivative

Table 7. RP4 Mediated Mobilisation of *Mob⁻* Vectors Derived from RSF1010.

Plasmid	Mobilising plasmid	Frequency of transfer per donor cell[a]
pKT230	–	$< 4.8 \times 10^{-7}$
pKT230	RP4	7.5×10^{1}
pKT231	–	$< 1.7 \times 10^{-7}$
pKT231	RP4	2.0×10^{1}
pKT262	RP4	7.2×10^{-8}
pKT263	RP4	3.7×10^{-6}

[a]*P. putida* KT2440 rif^r (rifampicin) was used as the recipient for transfer.

containing a *Tn3* element inserted into the *mob* locus was isolated. The plasmid was then linearised with *Bam*HI, the site for which is located in the *Tn3* element. It was then subjected to digestion with *Bal*31 exonuclease to form the plasmid pKT261. Both the *Tn3* element and the *mob* region are completely deleted from this plasmid. Vectors defective in mobilisation which are analogous to pKT230 and pKT231, namely pKT262 and pKT263 (*Table 2*) were then constructed. *Table 7* illustrates the effect of deleting this region of the RSF1010 genome on the ability of RP4 to mobilise the plasmids. In both cases the mobilisation frequency has been reduced by 5−6 orders of magnitude.

These plasmids, used in conjunction with the hosts *P. putida* 2440 and *P. aeruginosa* PAO1162 (9), constitute a host-vector system which fulfills the requirements for certified HV1 systems, as laid down by the U.S. Recombinant DNA Advisory Committee and the West German Central Commission for Biological Safety.

5. ACKNOWLEDGEMENTS

I would like to thank Professor M.Bagdasarian for his support in writing the manuscript, Elizabeth Badger for typing it, Susan Haley for drawing the diagrams and Bryn Price for preparing the photographs.

6. REFERENCES

1. Dalton,H. and Mortensen,L. (1972) *Bacteriol. Rev.*, **36**, 231.
2. Lundgren,D.G. and Silver,M. (1980) *Annu. Rev. Microbiol.*, **34**, 263.
3. Hütter,R., Nuesch,J. and Leisinger,T. (eds.) (1981) *Microbial Degradation of Xenobiotics and Recalcitrant Compounds,* published by Academic Press, London.
4. Franklin,F.C.H., Bagdasarian,M., Bagdasarian,M.M. and Timmis,K.N. (1981) *Proc. Natl. Acad. Sci. USA,* **78**, 7458.
5. Thomas,C.M. (1981) *Plasmid,* **5**, 10.
6. Don,R.H. and Pemberton,J.M. (1981) *J. Bacteriol.*, **145**, 681.
7. Guerry,P., van Embden,J. and Falkow,S. (1974) *J. Bacteriol.*, **117**, 619.
8. Watanabe,T., Furuse,C. and Sakaizum,S. (1968) *J. Bacteriol.*, **96**, 1791.
9. Bagdasarian,M., Lurz,R., Rückert,B., Franklin,F.C.H., Bagdasarian,M.M., Frey,J. and Timmis,K.N. (1981) *Gene,* **16**, 237.
10. Scherzinger,E., Bagdasarian,M.M., Scholz,P., Lurz,R., Rückert,B. and Bagdasarian,M. (1984) *Proc. Natl. Acad. Sci. USA,* **81**, 654.
11. Rubens,C., Heffron,F. and Falkow,S. (1976) *J. Bacteriol.*, **128**, 425.
12. Chang,A.C.Y. and Cohen,S.N. (1978) *J. Bacteriol.*, **134**, 24.
13. Bagdasarian,M. and Timmis,K.N. (1982) *Curr. Topics Microbiol. Immunol.*, **96**, 47.
14. Tait,R.C., Close,T.J., Lundquist,R.C., Hagiya,M., Rodriguez,L. and Kado,C.I. (1983) *Biotechnology,* **1**, 215.
15. Leemans,J., Langenakens,J., DeGreve,H., Deblaere,R., Van Montagu,M. and Schell,J. (1982) *Gene,* **19**, 361.
16. Clewell,D.B. and Helinski,D.R. (1969) *Proc. Natl. Acad. Sci. USA,* **62**, 1159.
17. Holmes,D.S. and Quigley,M. (1981) *Anal. Biochem.*, **114**, 193.
18. Birnboim,H.C. and Doly,J. (1979) *Nucleic Acids Res.*, **7**, 1513.
19. Maniatis,T., Fritsch,E.F. and Sambrook,J. (1982) *Molecular Cloning. A Laboratory Manual,* published by Cold Spring Harbor Laboratory Press, NY.
20. Haas,D. (1983) *Experientia,* **39**, 1199.
21. Holloway,B.W. (1965) *Virology,* **25**, 634.
22. Kushner,S.R. (1978) in *Genetic Engineering,* Boyer,H.W. and Nicosia,S. (eds.), Elsevier/North Holland, Amsterdam, p. 17.
23. Miller,J.H. (1972) *Experiments in Molecular Biology,* published by Cold Spring Harbor Laboratory Press, NY.

24. Boulnois,G.J. (1981) *Mol. Gen. Genet.,* **182**, 152.
25. McKenny,K., Shimatke,H., Court,D., Schmeissner,U., Brady,C. and Rosenberg,M. (1981) in *Gene Amplification and Analysis,* Vol. II, Chirikjian,J.C. and Papas,T.S. (eds.), Elsevier/North Holland, Amsterdam, p. 383.
26. Vieira,J. and Messing,J. (1982) *Gene,* **19**, 259.
27. Ish-Horowicz,D. and Burke,J.F. (1981) *Nucleic Acids Res.,* **9**, 2989.
28. Friedman,A.M., Long,S.R., Brown,S.E., Buikema,W.J. and Ausubel,F.M. (1982) *Gene,* **18**, 289.
29. Knauf,V.C. and Nester,E.W. (1982) *Plasmid,* **8**, 45.
30. Frey,J., Bagdasarian,M., Feiss,D., Franklin,F.C.H. and Deshusses,J. (1983) *Gene,* **24**, 299.
31. deBoer,H.A., Comstock,L.J. and Vasser,M. (1983) *Proc. Natl. Acad. Sci. USA,* **80**, 21.
32. Bagdasarian,M.M., Amann,E., Lurz,R., Rückert,B. and Bagdasarian,M. (1984) *Gene,* **26**, 273.
33. Sakaguchi,K. (1982) *Curr. Topics Microbiol. Immunol.,* **96**, 31.
34. Nakai,C., Kagamiyama,H., Nozaki,M., Nakazawa,T., Inouye,S., Ebina,Y. and Nakazawa,A. (1983) *J. Biol. Chem.,* **258**, 2923.
35. Stanisich,V.A., Bennett,P.M. and Richmond,M.H. (1977) *J. Bacteriol.,* **129**, 1227.
36. Casadaban,M.J., Chou,J. and Cohen,S.N. (1980) *J. Bacteriol.,* **143**, 971.
37. Hashimoto-Gotoh,T., Franklin,F.C.H., Nordheim,A. and Timmis,K.N. (1981) *Gene,* **16**, 227.